Chasing a Dream in the Galápagos

A Personal Evolution

Bette Blaydes Pegas

Sunbelt Publications
San Diego, California

Chasing a Dream in the Galápagos: A Personal Evolution
Sunbelt Publications, Inc.
Copyright © 2009 by Bette Blaydes Pegas
All rights reserved. First edition 2009

Cover design by Shelley Crutz, San Diego, California
Cover photograph: Gerald and Buff Corsi, (c) California
  Academy of Sciences
Interior design and layout by Robert Goodman, Silvercat™,
  San Diego, California
Printed in the United States of America

No part of this book may be reproduced, stored in a retrieval system, transmitted in any form or by any means, electronic, mechanical, audio recording or otherwise, without prior permission of the author and publisher. Please direct comments and inquiries to:

Sunbelt Publications, Inc.
P.O. Box 191126
San Diego, CA 92159-1126
(619) 258-4911; fax: (619) 258-4916

12 11 10 09   5 4 3 2 1

Library of Congress Cataloging-in-Publication Data

Pegas, Bette Blaydes.
  Chasing a dream in the Galapagos : a personal evolution / Bette Blaydes Pegas. -- 1st ed.
    p. cm.
  Includes bibliographical references.
  ISBN 978-0-916251-97-0
  1. Natural history--Galapagos Islands. 2. Galapagos Islands--Description and travel. 3. Pegas, Bette Blaydes--Travel--Galapagos Islands. 4. Pegas, Bette Blaydes--Family. 5. Natural selection--Galapagos Islands. 6. >Nature--Effect of human beings on--Galapagos Islands. I. Title.
  F3741.G2P44 2009
  918.66'502--dc22
                                        2009002034

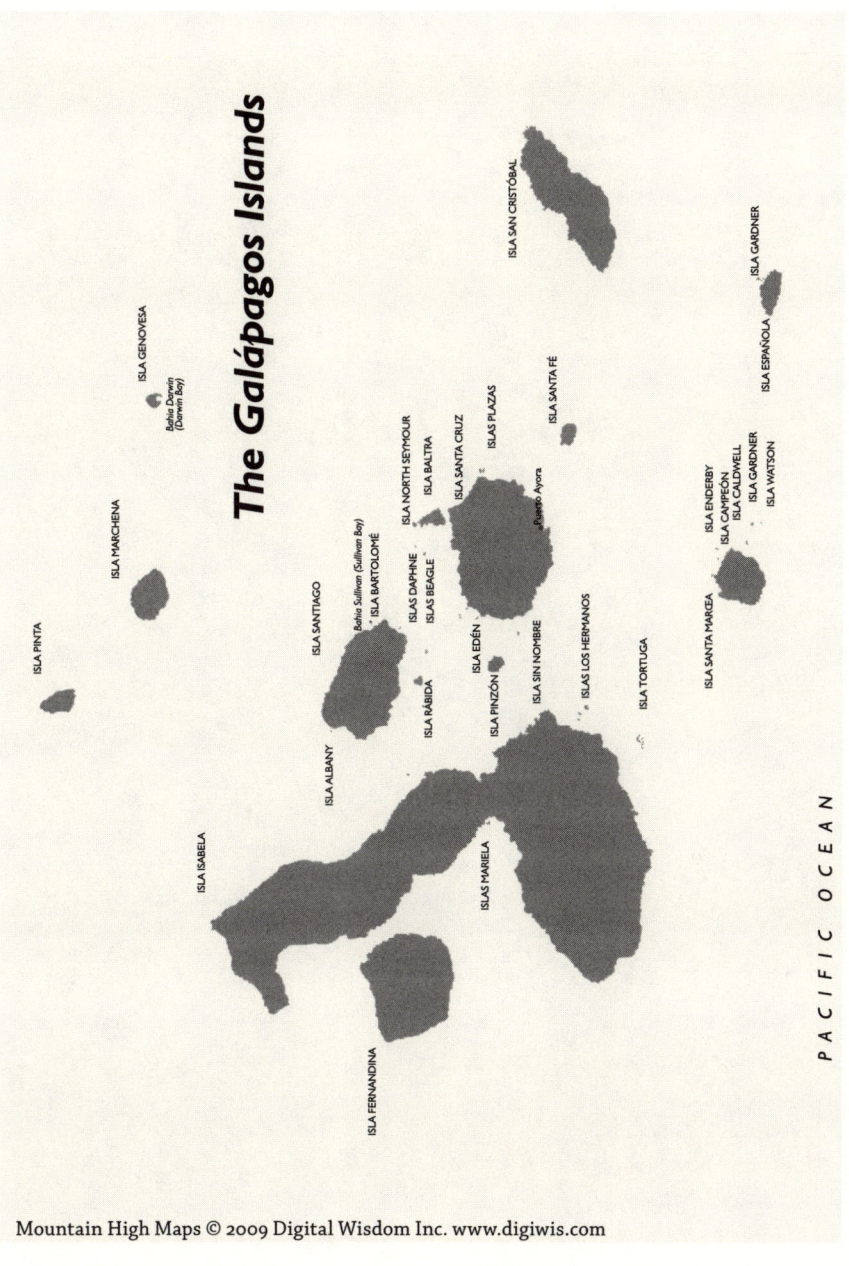

Note: Isla Darwin and Isla Wolf located northwest of the main Galápagos chain of islands are not included in this map.

To my mother-in-law, Katherine Mazarakis Pegas, who at ninety-nine has "never been happier"

And in memory of John and Celia Blaydes, my father and mother; Ralph and Estela Noriega, my uncle and aunt; my sister-in-law Connie Doukas, and my brother-in-law George Pegadiotes

Contents

Author's Note • ix
Acknowledgments • xi
1   The Question • 13
2   A Promise • 19
3   Discoveries in the Air • 25
4   Immersion • 33
5   A Growth Experience • 37
6   Palaces and Shoes • 43
7   Beyond Books • 47
8   The Dream • 53
9   Hearts in the Highlands • 57
10   Sights of the City • 65
11   Seaward Bound • 71
12   A Red Beach • 79
13   Snorkeling for the Meek • 85
14   God's Creatures and Lava • 89
15   Walls Apart • 97
16   Darwin Bay without Guidance • 103
17   Darwin Bay with Guidance • 107

18   Moon Landing   •   111
19   Discord in Paradise   •   117
20   Survival   •   121
21   "Valentine" Island   •   127
Epilogue: Full Circle   •   135
Chronicle of the Enchanted Isles   •   141
Discovering Charles Darwin   •   145
Bibliographies
Galápagos and Quito   •   163
Darwin   •   166

Author's Note

Because I was unable to contact my wonderful fellow passengers or the fine Galápagos guide and crew on the *Beluga*, I have chosen to change their names. I've taken the liberty of adding new information to the guide's narration and of distilling or blending the dialogue spoken by the passengers. All the names used in the descriptions of Quito are real. As with the island chapters, I have added new information to the narration of our unforgettable guide Nelson and reflected to the best of my ability our conversations. Memory is, after all, painfully fragile.

I hope that those who were with me and my family on that marvelous journey will recognize themselves and re-live this life-changing adventure. I have thought of each and every one of them throughout the eight years it has taken me to write this book.

Acknowledgements

To all of you who've listened to me talk about the "book I'm writing," here it is—for real—after eight years of promises. Often during these years, I've needed and received hundreds of "Don't-give-ups" and "We-know-you-can-do-its." Now, it's my turn to say thank you to those who have made this dream possible.

Let me start with my superb writing network and wonderful friends: Margaret Harmon, Dorothy Ledbetter, and Judy Barkley who've read, reviewed, revised and cheered me on month after month after month; and celebrated my upcoming publishing with a "literary lei." Why they didn't give up and scream, "Oh, no—not again!" I'll never understand.

Then there's my incredible family—my cousin-partner Yolanda Devlin (a.k.a. Yolis) who helped make the journey possible and has never stopped working throughout the entire publishing process. She is the sister I never had. I could not have survived without her. My husband, Art, a walking *Thesaurus*, never failed to give me the right word or phrase when I was struggling. Art's love and faith in me have kept me going. He stayed behind while I traveled and has cheered me on with his wry wit whenever I've been discouraged. My daughters, Heather and Suzanne, shared my dream journey and boosted my morale whenever I'd lose hope. When news came regarding agents or publishing, bad or good, I could count on them for consolation or cheers. And there's Yolis' husband,

Jack Devlin, with his ready smile and encouragement, who never complained when Yolis worked long hours with me and who fixed us luscious meals to keep up our energy.

And thanks to the special people who have either read my drafts, offered suggestions and support, or rooted for me non-stop—or all the above: Carole Hotelling; Martha Cummings; Ron, Gale, and Jay Vavra; Jan Crosby; Manny Skoor, Lori Baxter; Linda Negrete; Sherwood Holland; Teresa Lopez; and my cousins Raúl Noriega and Rosa María Noriega. And to Mike Merrill—always available to help when I had my many computer glitches.

Special professional thanks to Jennifer Redmond, the editor-in-chief for Sunbelt Publications Inc., who took a risk with this book and spurred me on to do things I didn't think were possible. To Robert Goodman, president of Silvercat, for his patience, professionalism, and expertise in assembling everything a book could need. To Shelley Crutz for her inspiring, artistic cover design and her spirit of helpfulness. To Peggy Lang, writer and editor, whose thoughtful criticism and direct questioning helped me refine what I wanted to say. To David Broad and the California Academy of Sciences whose photos have helped me tell my story.

When I began this memoir, I didn't realize how much help I would need. I am indebted to Dr. Linda Cayot, science advisor to the Galapagos Conservancy, for her painstaking review and recommendations on the scientific aspects of the book; and to Dr. David Archibald, evolutionary biology professor at San Diego State University, for his expert, generous assistance with the Darwin essay; to Johannah Barry, president of the Galapagos Conservancy and Lori Ulrich, marketing director for the conservancy, for their answers to my many island questions, their referrals, and for their support of my book.

1
The Question

Monday, March 13, 2000
El Cajon, California

Action is the flip side of dreaming, but I am tired today, too tired even for dreaming. No one is returning my calls, and the grant writing deadline is just days away. I impulsively, almost humorously, e-mail my daughter Heather, a fellow writer who understands my frustrations.

"Honey, I need to get away from all this. I think it's time for the Galápagos."

"What are you waiting for?" comes the reply minutes later. "Why not now? You'll never be younger or more able, Mom. Let's do it."

I stare at the message and remember how it all started more than twenty-five years ago.

"Shhh, Girls. Look at the bird on the TV." Heather and Suzanne stop their quibbling and look at the screen. A tiny, nondescript, yellowish-brown finch grasps a cactus spine with its bill to probe a pea-sized hole in a branch. He angles his head in the direction of the hole and pushes the spine downward, his eyes focused on the job to be done. Seconds later he spears an insect, pulls it from the hole and then stands on the spine with one foot while eating. I watch the

finch and block out everything else in our family room—my husband, Art, playing chess and our daughters squabbling over the ownership of a Chinese checkers game.

"This bird is smart, really smart, Girls. He knows his beak isn't long enough to reach down into that hole to get his goodies, and so he makes himself a tool." My eight- and six-year-old watch briefly and then run outside.

"The woodpecker finch," the narrator announces, "is found only on the Galápagos Islands. It is a rare bird that can make, modify and use tools and even carry tools around to reuse them. This finch is one of thirteen currently recognized Galápagos finches that contributed to Charles Darwin's discovery of natural selection. Natural selection, very simply put, is the process by which organisms better adapted to their environments leave more offspring than those less adapted. The more offspring, the greater chance for survival of the species."

Art stops his chess and together we watch other finches on other islands, each with a uniquely shaped beak. Darwin didn't know these birds were all from the finch family, the narrator informs us, until a London Museum ornithologist studied his specimens. The shape and size of their beaks, Darwin deduced, must depend on the food available to them on their particular islands.

If an island had lots of large nuts, Darwin reasoned that only the finches with powerful beaks to crack them would survive. If fruit-bearing plants flourished on an island, finches would need parrot-like beaks. And if insects were the major food source, the only ones to survive would be finches with slender, pointed bills. After much study, Darwin realized that if finches evolved into totally different species to *survive in their environments, so also must other living creatures and plants.*

A week after the Galápagos special I am washing dishes and thinking of Charles Darwin's powers of observation when I drop a chunk of celery down the drain. I try stabbing it with a knife, but find the knife too wide to reach it. Next I try an ice pick, but push the celery deeper into the drain. By the time Art arrives, I have subdivided and squashed it.

"I wonder how the woodpecker finch would have solved the problem," I think aloud.

"Still thinking about him, are you? He probably would have grabbed this very long, very narrow plastic spoon."

"Not a bad answer. You'd survive, but I wouldn't."

My husband laughs and scoops out the celery with his spoon tool while I continue to ponder the intelligence of the finch. How does the finch know the exact length and width of the cactus spine or twig needed to extract his grub, I wonder. Does he learn through trial and error the way we learn? How would my hero Charles Darwin answer my questions? Unlike me and my ilk who ask questions and then proceed with ordinary life, Darwin focused and studied and never gave up until he found answers.

Years pass and I don't forget the finch, the man, and those islands that formed beneath the sea millions of years ago. But I am so busy being a mother, wife, English teacher, and writer that going to the Galápagos becomes a "passive" obsession. Somehow, the dream itself fulfills me because it connects me with a world of noble creatures that live nowhere else. But only the adventurous, I tell myself, travel to Las Islas Encantadas.

One day I invite Jim, a science teacher friend, to show his Galápagos slides to our family. Jim has flown from Ecuador in a small plane, and toured the islands while based on a ship holding about ninety passengers. Each day, groups would head out from the ship on rubber dinghies or *pangas* and climb rocky cliffs under the equatorial sun to see booby birds, tortoises, sea lions, and marine iguanas. By late evening when Jim leaves our home, my dream has strengthened—but so also have the mental impediments to it, especially my fear of flying in small planes. I go to bed that night picturing myself seated behind the pilot in Jim's rinky-dink plane. I wake abruptly, clutching the imaginary armrests of a downward moving aircraft arrowing into an inky sea.

Art and the girls delight in hearing me talk about the trip, but seem to know that the talking is an emotional exercise. Besides, no one wants to travel with me to the home of Darwin's finches. "Where the sidewalk stops there also do I," Art chants. "Why the Galápagos?" Heather and Suzanne ask. "Why not Europe or Australia?"

In the early eighties, a travelogue about the islands comes to our town's fine arts auditorium. "How about seeing it with me?" I invite my daughters. We walk into a crowded theatre and sit in red velvet seats. Music billows from the speakers, and the huge screen fills with turquoise waves splashing on volcanic rocks. We watch tiny penguins—the color of hardened lava—strut and hop feet-first into the surf; fur seals swim among tourists in intimate, watery caves; and parades of marine iguanas fling themselves over rock ledges into the sea. After the presentation, my daughters and I hold hands on the way to the parking lot. "We want to go with you," they say. At that moment, my dream passes to them.

The Question

Throughout the years, we talk about how someday we will go together, but I never collect brochures, compare prices, call airlines, or talk with a travel agent. And whenever I read or hear of exotic South American diseases, or of tourists being accosted or killed in Quito, the Ecuadorian city we would need to visit before flying to the islands, I fear for my children. Dreaming will just have to do.

Sometimes we unknowingly bequeath our dreams to our children and then meet those dreams again when they come full circle. My daughter's words make perfect sense, but how can I fly to Ecuador *now* when my ninety-one-year-old mother-in-law is caring for my sister-in-law who is ill? And what about writing deadlines? And my frequent nosebleeds? My doctor may not consider them serious, but high altitudes can trigger them, and Quito is more than nine thousand feet above sea level.

For me, planning a trip of this magnitude would mean months of living in high gear. Oh sure, I'm used to multi-tasking, but when I'm over-extended, I forget to eat, then gorge on graham crackers, peanut butter sandwiches, and Wheat Thins®. I stay up long after midnight writing, balancing my checkbook, catching up with e-mails, and doing an exercise routine that I should have done in the morning. I'm not sure this trip would be worth what I'd do to myself.

Maybe I'm not meant to be a traveler. I grew up in a non-traveling family, the only child of parents who didn't own a car. My father kept his vow never to drive because as a teenager he'd almost run down a child in his friend's 1922 Chevrolet. The only trip I ever took with my father and mother was a train excursion to Los Angeles that abruptly ended when Dad became ill. A walk or bus ride downtown or to a neighborhood movie house was an adventure for my little family until the summer when I was almost seven and the Moynahans next door bought a chocolate-colored, four-door Hudson, with slippery seats long enough for six children or four adults. During evening drives with my best friends, Tim and Kathy, and our parents, I would sit with my nose pressed against the cool glass, eyes wide, scanning the path of the moon. "Does it go where I go or do I go where it goes?" I'd ask my parents who'd laugh and continue with talk that only they understood. Those Hudson evenings were my first taste of a world that traveling could bring.

Sitting at my keyboard, I think of the little girl in pursuit of the moon and of the woman who settles for the mundane. I'd need a passport, immunizations against God knows what. My Spanish is weak and my nerves taut. It would be so much easier to say, "No," and just let my daughters do it together. They could send me postcards and I could sit here and hate myself for being a coward because it started as my dream, damn it.

So I type "yes." And tremble.

2
A Promise

April 2000 through August 2000
El Cajon, California

"Artie, why don't you join us?" I teasingly, yet wistfully, ask my husband.

"Oh no you don't," he smiles. This is your dream. I'll feed the pets, water the trees, and worry about you until you're back with me."

I picture Art with his two titanium knees climbing in and out of *pangas*. Even without the knee problem, he'd rather play chess, bowl, read and write in his favorite coffee shops, or argue politics with his buddies. Like my father, Art prefers the comfortable and familiar.

A call to my social studies teacher-daughter Suzanne finds her anxious to join her sister and me. We then invite my cousin Yolanda, better known to us as "Yolis." Not only is she a risk-taker and fun to be with, but she also speaks fluent Spanish and used to be a travel guide. With Heather and Suzanne both in Oakland and Yolis and me in San Diego County, we communicate frequently by phone and through e-mail.

Our trip, we decide, must be in the middle of August so that Suzanne can be back before school starts. Starting in April, we contact dozens of travel agencies by phone and Internet. We agree that the most we can afford is five days on the islands, and that we don't want to be on a cruise ship with hundreds of people. Our boat must be more private, but not so small that we'll be storm-tossed and retching.

One agency after another turns us down. We want too much for too little. Our plans are too complicated. If all we wanted was a trip to the islands, they tell us, they could accommodate us easily. But three of us (Yolis, Heather, and I) are also going to the Inca city of Machu Pichu in Peru. We persist, using Internet contacts and phone numbers from travel brochures and articles.

In May and June, two agents tell us all is arranged and then cancel or quote prices way above our budgets. I tell Connie, my sister-in-law, yellow with jaundice, that the trip may never happen.

"Ridiculous. You are meant to take this trip, and you must take it." She recalls her own trip to Grand Canyon with her then-teenage daughter Valli and how they lay in sleeping bags on a truck bed watching the stars. "I want you to have what I had. Promise me you won't give up on your dream."

I stay afloat by daydreaming. I'm gliding on a raft in the eastern Pacific, six hundred miles off the coast of Ecuador. Glistening waves push me ever closer to a cluster of desert islands in a cold sea. Sometimes I see myself bounding through high grasses to commune with giant tortoises. Other times, I'm walking along a volcanic beach at sunset or watching one of Darwin's woodpecker finches spear a wriggling worm with a cactus spine.

In early July, Judy Martin from the Latin America Reservation Center (LARC) out of Florida calls to confirm our perfect package with Angermeyer's Enchanted Expeditions out of Ecuador: two days in Quito and five days on the islands. Then back to Quito for another day, followed by a flight home

A Promise

for Suzanne—and Peru for the rest of us. We will be based on the Beluga, a steel-hulled, one hundred ten foot yacht with a breadth of twenty-three feet at its widest point. It was built in Hamburg in 1968 and holds sixteen passengers.

Subject: Finally Coming Together
From: Yolis
Date: Mon, 10 Jul 2000

So, fellow adventurers, we are paid for and confirmed on everything. We're leaving August 16!! Our grand total for both Ecuador and Peru is finally in our ballpark. Unbelievable that it took us five months to get from too much money for a big cruise ship to "our price" for a private yacht, but it was worth it. Suz will fly from San Francisco to Houston. The rest of us will fly out of LA—even Heather. We'll meet Suz in Houston.

On Sunday morning July 17, I receive a panicky call from my mother-in-law. "Connie's very sick, and she's asking for you. Please come over as fast as you can. She doesn't want a doctor. She wants you." I page Art at his gym and drive to the Lemon Avenue home I've been popping into for more than thirty years. Traffic is light and I mindlessly maneuver my Toyota down winding, tree-lined roads. I've been dreading this moment for months, while watching her once-fair complexion turn amber, and her weight plummet from one hundred twenty to barely ninety. How I've missed "Con-Con," the clown, doing her Queen Elizabeth and Richard Nixon routines—the royal monarch waving her hand to an invisible crowd; the former president, jowls jiggling, as he flashes the victory sign. Gone are her idyllic descriptions of the emerald summers and crimson falls of her childhood in Pennsylvania; gone too those cozy, wee-hour kitchen powwows on books, family gossip, baking tips, and tales of horrific bathrooms we've visited. Lately, whenever I visit, she's been distracted and depressed. And when my mother-in-law and I urge her to see a doctor, she snaps: "Doctors can't help me." Just yesterday, she shuffled off to bed at three in the afternoon, leaving Mom and me alone and worried.

Within fifteen minutes, I am standing at Connie's bedside. Barely conscious, she takes my hand: "I knew you'd come." I kiss her damp forehead and glance up at my mother-in-law leaning against the wall to steady herself. When Art and his brother Manny and wife Marina arrive, the brothers cradle their sister in strong arms and drive her to the ER where she dies an hour later.

For weeks after Connie's death, I have to push myself to do anything. Struggling with my own grief, I think about Darwin and his many losses. Two of his ten children died in infancy. One child, Annie, died when she was ten. Darwin was forty-two when Annie became critically ill. His wife, Emma, was imminently expecting a child and couldn't travel to the dispensary where their daughter was being treated. Darwin rarely left Annie's side during the last ten days of her life and wrote Emma daily about her condition. A week after she died he wrote that he had "lost the joy" of his household. "She must have known how much we loved her; oh that she could now know how deeply, how tenderly we do still and shall ever love her dear joyous face."

Darwin buried his grief in scientific research. I bury mine in frantic activity—getting my passport and vaccinations; picking up prescriptions for altitude sickness and dysentery;

A Promise

and power-walking up a steep hill near our house to build endurance. I shop for an Ecuador-Peru guidebook, wrap-around sunglasses, comfortable shoes, notebooks, ten rolls of film, and a backpack with wheels. I find a floppy cotton camouflage hat in an army surplus store and crinkly tan pants with many pockets in the boy's department at Ross. I sign up for an anxiety group at Kaiser, and avoid news headlines on South American political unrest. My promise to Connie gives me motivation and energy.

Wednesday, August 16: No More Dreaming

I pop awake at the ungodly hour of five a.m., put on Art's many-pocketed western shirt with my own multi-pocketed pants, and stumble into the kitchen for a banana and hot tea. Art is dressed and watching the news.

"You're actually up on your own? I was just going in to wake you. I want to leave as soon as you're ready." Minutes later: "Are you ready yet?"

Driving to Yolis' house, my husband says very little for the first few miles. Speeding through the light freeway traffic, he tightens his grip on the steering wheel. I watch the muscles of his strong arms tense and his knuckles, like outcroppings, stand out against the surface of his hands. When an AP bulletin on the radio announces that a burst tire was the "primary cause" of last month's Air France Concorde crash, he switches to country music. I fumble through my pockets, checking to see which one holds my passport and tickets.

About three-quarters of the way to our destination, he reaches for my hand across the seat. "I see you're wearing my shirt. You'd better take care of it."

"I love this old shirt. I guess I'm kind of taking you with me—way past where the sidewalks stop."

"Bette, I know you'll be fine, but if anything happens—If you need me, I'll get to you no matter what—sidewalks or not."

I rub the back of his neck. "I know."

In front of my cousin's home, we say our goodbyes and I promise to "be careful." Yolis' husband, Jack, then drives us one hundred fifty miles to LAX where we meet Heather. After a three-hour plane trip to Houston, we connect with Suzanne.

Voices echo and merge in the huge Houston International Airport. I follow my daughters and Yolis at a fast clip to the area where a 737 Continental Airlines jet awaits to fly us to Quito. With the daypack on my back and pulling the other bag behind me, I am off-balance and a bit giddy. I pat myself down with my free left hand. The money belt is secure under my shirt. I feel for my passport, double-check my many pockets for tickets, cash, and my copy of Darwin's *Voyage of the Beagle*. After boarding, we search for our seats, hurling backpacks into the carriers above us, and scooting daypacks under our seats.

As we lift off the ground, I close my eyes and invoke protection for us all. I think of Connie and a calmness settles over me. My trip had resonated with her, brought back feelings she hadn't spoken of in years. Her words still linger: "I want you to have what I had." Sitting with my family and listening to my breath flow in and out, I touch my Darwin and feel a tear gain momentum as it zigzags down my cheek.

3
Discoveries in the Air

Wednesday, August 16
En Route to Quito, Ecuador

Sometimes I shock myself. In all my preparations and years of dreaming, I have never taken time to read Darwin's own words about his journey to South America. I maneuver my cramped legs into a "perfect" position—one that will bring comfort for at least ten minutes—place the *Beagle* diary on the pullout tray, and open to the chapter on Galápagos.

I struggle to replace my customary image of the elderly, bearded scientific genius with the picture of a brilliant young man writing the adventure story of a lifetime. Darwin yearned to be a part of the Beagle expedition, but didn't want to disappoint his father. Strong-minded Robert Darwin insisted that Charles become an Anglican minister and that taking such a trip would divert him from his true goal. Darwin's voyage never would have taken place had his uncle not interceded for him. His father finally yielded and gave him permission to sign on for nearly five years as an unpaid naturalist on the *HMS Beagle*. According to Darwin, the purpose for the expedition was to survey Patagonia and Tierra del Fuego, the shores of Chile, Peru, and "some islands in the Pacific." When he reached the Galápagos in 1835, he was only twenty-six and had already collected hundreds of animal and plant specimens and fossil remains.

I read with amazement his zoological, botanical, and geological observations, and his almost innocent reflections as he begins to assemble a giant scientific puzzle:

> The natural history of these islands is eminently curious and well deserves attention. Most of the organic productions are aboriginal creations, found nowhere else; there is even a difference between the inhabitants of the different islands, yet all show a marked relationship with those of America.... Hence, both in space and time, we seem to be brought somewhat near to...that mystery of mysteries—the first appearance of new beings on earth.

Reading on, I have to remind myself that Darwin lived more than a hundred fifty years ago—long before modern ecological awareness and endangered species protection. And even though I know I'm being irrational (He is, after all, hungry and has few diet options), I find myself shocked and disappointed when he eats and bedevils creatures that are now shielded and even venerated.

On October 8, 1835, he writes: "We arrived at James Island.... While staying in this upper region, we lived entirely upon tortoise-meat: the breast-plate roasted...with the flesh on it, is very good; and the young tortoises make excellent soup; but otherwise the meat to my taste is indifferent."

"Not only does he eat tortoises—he teases iguanas," I tell my family. "Here in his journal, he describes a land iguana digging a burrow: 'I watched one for a long time, till half its body was buried; I then walked up and pulled it by the tail;

Charles Darwin at thirty-one

at this it was greatly astonished, and soon shuffled up to see what was the matter...."

Suzanne blurts out her surprise. "The great Darwin pulling iguana tails? It pains me, but I need to read that when you finish, Mom." Yolis and Heather look up from their books.

"Listen to what he says about marine iguanas," I continue:

> They do not seem to have any notion of biting; but when much frightened they squirt a drop of fluid from each nostril. I threw one several times as far as I could, into a deep pool left by the retiring tide; but it invariably returned in a direct line to the spot where I stood.

"Unbelievable!" my family gasps almost in unison. I promise to pass Darwin's journal around when I've finished the "Galapagos Archipelago" chapter, and Heather promises to share *The Beak of the Finch*, by Jonathan Weiner.

"This book's a must-read," she tells us, "It actually makes evolution understandable and explains how life began on the Galápagos."

Not wanting to miss out on Heather's discovery, Suzanne stands and peaks over her sister's head while Yolis leans across the aisle, and I slide as close as I can to my cousin.

"Darwin believed that plants were the first life forms on the islands. He knew this because without plants to eat, land animals couldn't have survived on the barren lava soil. He figured there were two ways plants could reach the islands."

Heather explains how seeds could either have floated to the islands on ocean currents or been dropped on land by birds. To prove his ocean current theory, Darwin soaked seeds in briny water for forty-two days (the exact number of days he estimated it would take for them to float from the South American mainland to the Galápagos). When they sprouted, he knew he was onto something big.

"To check out his bird droppings theory, he fed oats to sparrows, fed the sparrows to a snowy owl and an eagle at the London Zoo, and waited for the owl and eagle to poop. He planted the poop, and when a seed sprouted, he knew he was right about how plant life and animals first came to the islands.

"But of course," we all agree. It seems so simple, so obvious when we hear it explained.

My daughters and cousin go back to their reading. But I can't stop thinking of Darwin's mind. When he was a child he germinated seeds in his father's gardens, and when his brother, Erasmus, built a chemistry lab in the backyard, the two would mix chemicals and produce gases his sister feared would blow up their house. When he went to Cambridge, instead of focusing on the ministry, he obsessively collected beetles. One day he discovered two rare beetles and grabbed one with his left hand and the other with his right. When he spotted a third, he popped the one in his right hand into his mouth, but it was so caustic tasting that he spit it out and lost it along with the third one.

We can snicker all we want about Darwin and the iguanas. But the truth is he wasn't just teasing, he was collecting information. I put down my book and watch my family in silence: Heather's ivory profile, Suzanne's dark braid curving over her shoulder, Yolis and her shiny-straight bangs. It doesn't seem very long ago when I watched seven-year-old Yolis organizing a fool-proof marketing scheme for breaking the neighborhood lemonade sales record. Or when I lowered my head into my first-born Heather's bassinette to check her breathing, so afraid I might lose her to Sudden Infant Death Syndrome as had my friend Pat with her baby. Or when I'd wake to a sound in the middle of the night and discover fifteen-month-old Suzanne asleep on her knees in the bathroom—her forehead pressed into the fuzzy green rug, her diapered bottom uplifted. For the first time in years I am with my girls and cousin for an unbreakable period of time. I am content. We fill the five-hour flight to Quito with philosophical ruminations and silly talk.

About three-quarters through the flight, Yolis holds out her left hand. "Hey you guys, how do you like my thrift-store wedding ring? I wasn't about to bring my real one." I look at my own left hand stripped bare of its diamond solitaire—now resting in its velvet box at home. I have only a simple white-gold band and a western shirt to remind me of my husband. He's worried, I know, and will probably panic every time the phone rings, thinking that something may have happened to us. I remember what he said in the car: "If you need me, I'll get to you no matter what."

Funny, I think. He doesn't even have a passport. Still, if he says he'd come, I know he'd find a way. I recall a long-ago night when I got locked in a bathroom some distance from the main building of a local high school. When I didn't show up for our date after my night class, he drove to the school and convinced a custodian that his fiancé was somewhere in the school and had to be found. I had almost given up pounding and yelling and was hunkering down for the night when I heard one of the sweetest sounds I've ever heard: "Bette, for God's sake, where are you?"

Heather breaks free from her *Finch* book: "Mom, you have to read this. Darwin was right about how creatures change and adapt, but he was wrong about how long it takes. It's not rare and slow. It's happening all the time, and his finches prove it."

I think again of the young Darwin. He had no idea he would revolutionize science and leave an indelible imprint on Western philosophy. He didn't even mention the Galápagos in the last chapter of his *Beagle* diary when he described what most impressed him about his journey. I flip through the pages of his journal and re-read his contemplations on the variety of finch beaks. "One might really fancy," he wrote, "that from an original paucity of birds in this archipelago, one species had been taken and modified for different ends." I visualize my favorite woodpecker finch digging grub with a tool he fabricated through his own ingenuity.

Seat belt lights begin flashing. "Ladies and Gentlemen, raise your food trays and fasten your belts. We will be landing in the capital city of Quito, altitude nine thousand four hundred forty-six feet. Temperature: sixty degrees. Time: Eleven p.m.—when the residents start partying. Notice the volcano Mount Pichincha on your left. It's still fuming from its last eruption in 1999."

Out the window, I see a cinnamon-red volcanic glow. Lit by an equatorial moon, the city awaits in the shadow of the Andes. My ears pop as we drop down, down into a basin of twinkling lights flowing from the center of the earth.

4

Immersion

Wednesday, August 16
Late Evening in Quito

Necesitas taxi? ... ¿Quieres ayuda? ... ¿Limpio sus zapatos? The airport is alive with questions.

My Mexican mother would smile to see me straining to re-familiarize myself with the words, phrases, and cadences of her beautiful language. For years I ignored her pleas to learn Spanish because I wanted to fit in with my all-Anglo friends. With my medium-light complexion and English-German father, fitting in was relatively easy, but costly. Only when my mother and I visited my Mexican family in El Paso and Juárez did I begin to understand what I had lost. My mother is dead now after more than eleven years with Alzheimer's. Toward the end she slipped back into her native tongue, and I struggled to make "her" language my own. It's been almost three years since I've used the language with any regularity.

Bienvenido a Quito. "Welcome." *¿Necesitas hotel?* "Do you need a hotel?" *¿Necesitas comida?* "Food?" *¿Necesitas cambio?* "Money changed?"

I'm in a listening stage now and can't speak. In perfect Spanish, Yolis explains that we already have accommodations and are looking for our guide from Angermeyer's. Minutes later he appears.

"Welcome to Quito, Señoras. My name is Nelson. I am your guide," he smiles. " I will wait while you go through customs. Don't worry. It'll be quick."

Less than an hour later we're riding through New Town as it vibrates with the laughter of youth and the sounds of salsa music booming out of wall-to-wall clubs or *salsatecas*. Taxis honk and our driver weaves through cars parked casually in the middle of narrow streets. In another twenty minutes, we pull up to a high white wall partially concealing the Angermeyer's Enchanted Expeditions office and the Orange Guest House behind it. A smiling guard opens the wrought iron gate, and we follow Nelson and the driver who carry our baggage through a courtyard and up an outside brick staircase. We pass an enclosure where the night employee, a man in his seventies, welcomes us and gives us our keys. We are on the second story—with two rooms next to each other.

"Be careful," Nelson tells us before he leaves. "You are four women alone. Your hotel area is safe in the daytime, but be sure to take taxis at night. The drug dealers and thieves busy themselves after sundown. And be sure *not* to wear jewelry. Have a good night. Oh, and don't overdo your activity. The altitude can make you dizzy. We'll be back for you at nine a.m. for a tour of Old Town."

Fleeting thoughts of drug dealers and thieves vanish as we reach our rooms. As for dizziness, we're in no condition to overdo activity. Heather and Suzanne stumble into their room while Yolis and I drag our bags next door. We take in our winter-cold surroundings—small, clean, with beamed ceilings and a barred skylight. Plopping down onto twin beds covered with blue sleeping bags, we unpack a few necessities.

My feet go numb as I step onto the cold bathroom tile. In the mirror, I stare at my eyes—like slits in piecrust—and splash icy water onto my face. I mustn't swallow the water, not even get drops on my toothbrush. I must use the bottled water provided. And what's this sign next to the toilet? *Papel no. Pónga el papel en la cesta.* I must remember to put paper in the basket—not the toilet.

Thursday, August 17—Upon Rising

I wake to the sounds of traffic, horns, and voices. It's about seven or eight. Who knows? Who cares? Beams of light pierce through me from the skylight fifteen feet above my bed. Warm in my sleeping bag, I feel the mountain chill on my face and remember the note in my guidebook about "four seasons in a day"—spring-like mornings, afternoons like summer, autumnal evenings, and winter nights. I've slept in my jeans and a long-sleeved shirt partly for warmth, and partly because I was sure we'd oversleep. Raising my head, I feel a warm trickle from my nose. Oh my God. It's going to happen. I'm going to hemorrhage. "Don't be ridiculous," I tell myself. "Just hold your head straight and take a deep breath." I brush my hand across the dampness and inspect one scarlet-stained finger. "Now, pinch your nose and be calm," I chant in a soothing voice. "You're going to be fine. Just be calm." In about five minutes the bleeding stops, but I know I must be careful.

Yolis is still sleeping, and so I shuffle quietly into the tiny bathroom, reminding myself about the nine thousand foot elevation. I must keep my head up and not bend over the sink which is way below my waist. Neck stiff, head erect, I fill my cupped hands with frigid water and let it drip gently into my eyes and cheeks, careful not to touch my nose. I use the bottled water to brush my teeth and almost break the "no paper in the toilet rule." I'm barely out of the bathroom and Yolis is already up, dressed and out the door.

¿Cómo pasó la noche? "How was your night?" Marcía, the daytime proprietor, asks when I enter the dining room. *Muy bien*, I tell her. *Dormí perfectamente.* "I slept perfectly." *Pero me siento un poquito tarda.* "But I feel a little sluggish." *Muy normal*, she responds, and gestures toward the continental breakfast buffet she has prepared with pride.

Minutes later at our table, we compare symptoms: "I don't feel dizzy, just a little unsteady," Suzanne says. Marcía reassures us that the feeling will pass as we adjust to the altitude. I say nothing about the nosebleed. After partaking of sweet rolls, papaya juice, and tea, we feel better. We thank Marcía for her hospitality and ask her to store some extra things we don't want to take with us tomorrow to the Galápagos.

Por supuesto. "Of course." *Van a estar seguros hasta que regresen.* "They will be safe until your return."

5
A Growth Experience

Thursday, August 17
Early Morning in Quito

As promised, Nelson and the driver arrive at nine. A native of Quito, Nelson tells us that his grandfather, a fan of naval stories, named him for the British Admiral Lord Horatio Nelson of *Hornblower* fame. Our Nelson is clean-shaven, in his early forties, slender—eyes bright with interest and respect. From his seat at the front of the van, he engages us in morning conversation. Did we sleep well? Was it too noisy? Are we ready for our tour? Did we know that our Secretary of State, Madeleine Albright, was in Quito now?

"The Ecuadorian people are quite impressed with her," he tells us.

"And President Clinton?" we ask.

"Yes, with him too. We hear about the scandals but are more interested in his policies."

He looks straight into our eyes. All of our questions seem worthy of a thoughtful response.

"I've heard that visitors can see inmates in the female prison on Thursdays," Suzanne, an activist and former social worker, interjects.

"I haven't heard about the female prison," Nelson tells her, "but the government will soon release all prisoners who are sixty-five years old and older. It will also release those who are sick—especially the terminally ill—and mothers. The prisons here are so crowded there is no way to contain them all."

I say nothing to Suzanne, but am ecstatic that we won't be visiting the prison. Of all the places I would not like to visit on this trip, the prison is definitely at the top of my list. I wonder what Nelson is thinking. I doubt he's ever been asked such a question and feel proud that my daughter, even on vacation, holds tight to her social conscience.

My schoolteacher persona ignites when I see children in uniforms carrying books and chasing each other down the street. "Are the schools here year-round?" I ask.

"Those children are from a public school. We have just finished a teacher strike for wages," he says. "To make up for the time lost, the public school children must attend school now, while private school children are on vacation."

I take out my blue mini-spiral notebook and scribble as we drive along. Others may take pictures. I take notes. We are deep in a valley surrounded by dark-green mountains at the foot of the smoking Mount Pichincha. From a lookout point, we survey white church steeples and hillsides dotted with houses that look impossible to reach. As we reach *El Parque Ejido* (the public park), we wave to students studying on the lawn and pass the parliament building where the lawmakers are currently in session. We laugh when Nelson refers to the building as "la casa de brujas" (the house of witches). His take on politicians is no different from ours. Entering Old Quito, we bump along blocks and blocks of cobblestone bordered with colonial architecture dating back to the sixteenth century.

We drive past a theatre named for national hero Mariscal Antonio José de Sucre and alongside bustling open plazas, once the gathering places for Spanish conquistadors, now teeming with tourists and native Ecuadorians eating, talking, and resting. A rush of baroque cathedrals, public buildings, monasteries, and museums floods my thoughts. I ride along in a delicious, visual fog unable to keep up with or make sense of all I'm seeing and not really caring because it's all so extraordinary. Occasionally, I hear something that brings me back to consciousness.

"Our city was named a World Heritage site in 1978," Nelson tells us, "because it has the best-preserved, least altered historic center in all of Latin America. What you are now seeing is protected by the United Nations. No changes can be made to buildings. They are just as they were in colonial days. No skyscrapers, just grand churches, about eighty-six of them, and government palaces."

Street vendors move among the tourists while old women and children, their hands extended, hope for riches to bring home after sundown. The smell of diesel mixes with the odor of baking bread in *panaderías* or bakeries, a vivid reminder of those rare summers when I'd cross the border into Juárez with my mother, grandmother, aunt, and cousins. My father never went with us on those trips. His employers at Boulevard Photo knew his value, knew they'd lose high-paying clients if Dad wasn't there to process film, make prints, and work his magic retouching portraits. "I can't go with you," he'd tell us. "What if that letter announcing my million dollar inheritance from Uncle Ben arrives and I'm not here to read it?" Mama and I would laugh. Mythical Uncle Ben had surfaced again.

"The *Iglesia de San Agustín* where Ecuador's Declaration of Independence was signed in 1809 is closed now," Nelson announces, "but we will get out at this corner to view its

A Narrow Street in Quito

external architecture. He reminds us to take off all jewelry, even costume pieces because they could attract thieves. "And make sure you have nothing dangling." Even the thrift store beaded chain holding my glasses could be a temptation, he tells me, and so I slip it under my shirt.

Suzanne asks about her multiple earrings: "Would it be all right to cover my ears with my hair?" He nods but looks worried.

We get out of the van. People throng. In seconds, a man about five-feet-six inches—slick black hair, bright white shirt—darts from across the street and lunges at Yolis, his fingers reaching for something beneath the high neckline of her shirt. She screams and just as his fingers grasp a now-visible gold chain and cross, she raises both arms and rams her knee into his groin. He groans and runs. Nelson is at her side immediately offering assistance.

"It happened so fast. I could not stop him. I am so sorry."

"It's my fault," she tells him. "I didn't remember the chain. Jack gave it to me for our wedding. I never take it off." She takes it off now. She is bleeding from a gouge in her chin made by the thief's fingernails and a second cut below her throat where the cross lies. "I'm glad it happened to me and not you guys," she tells us. "I've been through this kind of thing before." She dabs her cuts with a tissue. "Our husbands don't need to hear about this until we're home," she whispers.

We encircle my cousin—all of us shaky, but determined not to let this happen again. Suzanne removes all her earrings. I put the eyeglass chain into my pocket. Heather grips the money belt under her white shirt. And Nelson hovers even closer than before.

6
Palaces and Shoes

Thursday, August 17
Late Morning in Quito

When I hear of violence striking others, I ask myself what I would have done in the same situation. Today for the first time in my life I've witnessed an assault on someone I love. Though trivial in contrast to violence I read about daily, this attack is personal. The once abstract question is now real, yet still unanswerable. My little cousin doesn't tell us how frightened she is, but I feel her fear as I walk beside her. At intervals, she dabs the blood on her chin with tissue and talks bravely about the incident. I ponder my own possible responses. Would I have run away screaming? Would I have surrendered the symbolic cross? Or would I have pummeled the accoster with my fists and sent him running into the crowd to escape my wrath?

"There was no way I was going to let him get that cross," Yolis tells us as we climb back into the van and drive close to *La Cueva del Oso*, or "The Bear's Cave," a famous restaurant once a gathering place for Quito's bohemian set. Nelson watches us carefully as we leave the van again to tour the *Plaza de la Independencia*, the main square of Old Town Quito. Walking over cobblestones among floral gardens, we pass old men lounging in the shade of palms and pines. As we

Chasing a Dream in the Galápagos

Virgen de Quito, Guardian of the City

absorb the Ecuadorian sunshine, I take my cousin's hand. She seems all right, but I know her too well to be fooled by her composure.

One after another, three or four children with brushes and polish approach us. No school uniforms, no books among them. *¿Puedo limpiar sus zapatos?* "Can I clean your shoes?" With each request comes our standard answer: *Muchas gracias, no.* Up the stairs we climb to the presidential palace guarded by soldiers in high black boots and traditional uniforms of blue and white trimmed in red and gold. We watch the changing of the guard and listen to our own footsteps and voices rebounding off colonial walls, pillars, and tiles.

Back in the van, we head up *El Panecillo* or "The Bread Roll," a hill overlooking the city. At its summit is an enormous cast-aluminum statue of the winged *Virgen de Quito* protecting all who visit. Smiling boys offer to shine my shoes, but again I turn them down—not for lack of need (one look at my dusty black shoes shows extreme need) but because I don't want to start anything. I know I'm a soft touch, and once I start saying "yes," I'll feel guilty every time I have to turn down a pleading face.

With all my restraint, I'm not prepared for Oscar—the magic boy whose vibrant eyes make me vacillate and reconsider. Nelson is the first to employ the eight-year-old's labor. Maybe they know each other. Maybe this has been pre-arranged. Anyway, when Oscar approaches me: *Señora, yo puedo limpiar sus zapatos muy bien.* I can't resist. He tells me he's one of five children and wants to be a mechanical engineer. He brushes the dust from my shoes, then pulls out a clean brush which he scratches through his polish container—careful to take just enough—not too much. I inhale the polish as he applies it gently first to one shoe, then the other. He allows time for drying—then, with a soft cloth, massages them until they gleam. I ask Oscar how much it costs for all the polish in his wooden box. One dollar a week, he tells me, and so I give him two. *Muchas gracias, Señora. Voy a resar por ti.* I nod a thank you. I appreciate his prayers.

By one p.m. we're back in front of the Orange Guest House. I frantically search through my purse, my many pockets and under the seat cushion, but can't find my blue notebook. I tell Nelson and he asks the driver to search the van. Maybe it fell when we got out. Maybe it slid under the seat. No luck. He maneuvers under the passenger seat to see for himself.

"I am so sorry," he says. "Maybe you will still find it. Was it very important?"

"Not really," I lie. "Just my notes—my memories of Quito." He looks at me knowingly and checks once more under the seat.

"We will look again," he promises. "Maybe it'll turn up in your things. Until tomorrow," he waves. "I will be here at seven in the morning to take you to the airport for your trip to the Galápagos."

"How small will the plane be?" I ask.

"You mustn't worry, Señora. It'll be fine."

7
Beyond Books

*Thursday, August 17
Afternoon and Evening in Quito*

Thoughts of small planes and the blue notebook fade in a flurry of shopping for wet suits and dining on creamy potato soup and stuffed avocado. By three-thirty, we're on our way to *Mitad del Mund*—the middle of the world. Our cab driver, Miguel, has also promised to show us a volcanic crater village—just a little bit out of the way.

Outside the congested city, we whiz through an expanse of pine and eucalyptus and then pass through the tollgate to *San Antonio de Pichincha*. I forget where I am for a moment when I catch sight of a small girl skipping along in a circular plaid skirt and clinging to her mother's hand. Years ago, thousands of miles away on the streets of San Diego, I was that child in plaid holding tight to my own mother's hand.

Onward we ride, passing vineyards, cacti, and children on go-carts. Because Miguel speaks little English, Yolis becomes our official interpreter. "I hope you'll be able to see the crater village," she translates. "Usually, at this time (around four-thirty), there is a thick cloud cover." Miguel parks his cab about thirty feet from the crater and we walk against the cool wind to the edge.

I stand on the precipice of a village about a mile below me—looking down through crystal clear air to minuscule green plots of farmland. More than a hundred people live in the crater of *Pululahua* which last erupted four thousand years ago. The people support themselves, Miguel tells us, by raising corn, potatoes, wheat, and livestock. Many are descendents of the native *Huasipungeros*, the original settlers. I see tiny people walking down the steep sides of the crater to enter the village. On the opposite side is a road where cars and trucks may approach and exit. Most of the residents, however, don't have motorized vehicles—they live an almost Amish existence removed from the rest of humanity.

The developing cloud cover prevents us from adventuring farther than the brink on which we stand. "It would take at least a half-hour to hike down, and another hour to hike up," he tells us. I am strangely drawn to life in this village and wonder about the people who choose to live here. Though tired, I regret that my "visit" must be long-distance. Miguel points in the direction of a church steeple. "It is a Catholic village," he says. "And if you look over there, you'll see their school. They even have their own herbalist doctors, but in case of emergency can be helicoptered out."

Through the rapidly condensing vapors, I strain a final look at dozens of cattle and sheep balancing on the nearly vertical hillsides. "Look fast," he tells us. "In minutes it will all be gone." We stand in silence as the wind redistributes the final whispers of mist. Then, like the legendary Brigadoon, the village disappears.

Inside the cab again, I take out a fresh, red notebook. Holding pen tight against the page to compensate for potholes and sharp turns, I write my impressions. I find myself as interested in Miguel as in the sights blurring past us.

Beyond Books

Having agreed to a total cost for the afternoon, Miguel has nothing to gain from extending our visit. His desire to share, coupled with our respect for all he offers, brings us unexpected pleasures and insights.

"Our people are unified for freedom," he says. "We have had hard times here, but are making real progress."

When we comment on the lack of graffiti, he talks about an organization of indigenous or native people. "We do not like graffiti because it is a means of communication for the powerless, and in Ecuador—unlike in other countries—indigenous people have power." He describes how the taxi union—of which he is a member—sided with the indigenous people and shut down the city for days when the government levied an unfair gasoline tax. "Now, though gasoline is still expensive, it is at least affordable," he stresses.

We reach our destination by five-thirty and follow Miguel through a sprawling tourist "village" to an obelisk aligned with the points of the compass. He talks about a scientific conference held here in the 1500s to determine the exact location for the middle of the world. Like millions before us, we each straddle the red line that *symbolizes* the center of the earth—putting one foot in the northern hemisphere—the other in the southern. We then take an elevator to the top of a museum building overlooking snow-capped *Cotopaxi*, the second highest active volcano in the world following Chile's *Tupungato*.

Exhausted, we follow his brisk steps into the museum, walking down several flights of stairs through life-like displays of indigenous history. "For more than five centuries Indian people in Ecuador have struggled against exploitation and racism. What we want is a world that respects diversity." He points to two Otavaleño figures: a man in a black fedora, his long braid or *shimba* hanging down his back, his azure poncho and white breeches reaching to his calf; and a woman wearing a white peasant blouse, blue skirt and shawl, with golden beads around her neck. "The Otavaleño tribe is the only one to maintain purity of customs and traditions," Miguel tells us.

"Most Otavaleños speak Quichua and wear their native dress. They are famous for their textiles, especially their beautiful blankets, ponchos, and Panama hats which most people mistakenly think originated in Panama." Sensing his pride and excitement as we explore the museum, I feel embarrassed at my own lack of energy. We stop in front of an exhibit of Afro-Equadorians and learn that Africans first came to Ecuador in the sixteenth century on a Spanish slave ship that ran aground off the coast. The slaves overpowered the whites and established their own colony. Like the indigenous peoples of Ecuador, Afro-Equadorians continue to struggle for equal rights. Onward we float through colorful artifacts and housing replicas. Barely able to walk now, we tell Miguel how much we appreciate everything we're seeing, but must go to dinner and rest.

Driving us to *La Ronda*, a restaurant recommended by Nelson, Miguel promises to meet us outside when we are ready to leave. Nearly two hours later, we find him standing next to his cab under a street lamp. "Thank you, Señoras, for everything," he tells us when we arrive at the Orange Guest House. "Have a wonderful time in the Galápagos, and phone me at my home when you get back. I would love to show you more."

As I reach into one of my secret pockets to add to our generous tip, my fingers graze the spiral edges of a notebook. Pulling it out, I recognize my blue journal. I'd convinced myself, or so I thought, that its loss had been petty, not worthy of special attention—that it didn't matter, that I could easily recapture my memories. My joy and relief over its recovery surprise me. *Gracias, Miguel,* I tell him, holding up the found treasure. *Tú me has traido buena suerte.* "You have brought me good luck." I tuck money into his hand and promise to call on our return trip to Quito.

It is after midnight in the Orange Guest House. Yolis and I pack and repack, determined to take only the basics to the islands. We stuff unneeded clothing and cosmetic supplies into plastic bags we will entrust to Marcía until we return. Right now, we must sleep. In spite of music and laughter from the partying Ecuadorians in the club behind our guesthouse, we are comatose within minutes.

Hours later, a jet flies over and I startle awake, staring up through the skylight at the new day. I check my alarm watch which reads six-forty A.M. "My God!" I scream to a sleeping Yolis, then bang on the wall to awaken Heather and Suzanne. "Wake up! Wake up, everyone! I must have set my alarm for P.M. Nelson will be here in twenty minutes!"

8
The Dream

Friday, August 18
Early Morning on Baltra, a Central Island

Our meticulous packing, plus our decision to wear travel clothes to bed, pays off. With swollen eyes and barely combed hair, we all greet Nelson and the driver when they arrive at seven *puntual* to take us to the Quito airport.

"You found your notebook," he smiles as I hold it out for him to see. "I am so very pleased for you, Señora."

The drive between the guesthouse and the airport seems shorter than when we arrived only a day ago. And when we get to the airport, we are together with our friend Nelson who guides us in and out of lines and entertains us with stories of his experiences as a Galápagos guide years ago. Nelson stays with us until we re-enter the sunlight to climb the metal stairs to our plane. "Make beautiful memories, Señoras. I will be here when you return," he waves.

"If only we could take him with us," I tell my family and they agree.

We fly to Guayaquil, change planes—then on to Baltra, the gateway to the Galápagos. Looking out the plane window, I think back on years of excuses. Violence? Yes, but we got

Four Travelers (from left to right: Heather, Yolanda, Bette, and Suzanne)

past it. Small planes? None—I'm flying in a Boeing 727. Nose bleeds? Only one, and it was over fast.

Less than an hour later, the wheels bounce on the tarmac and my heart races. If Connie could only see me now. I am here. I am really here. I gather my things and press my way into an intimate line of bodies inching through the plane. I clank in slow motion down the metal staircase onto a sandy runway. The air is pure. The warm breeze plays with my hat, but the strap holds tight beneath my chin. We pose for pictures in front of a small cactus garden and a *"Bienvenidos a Galápagos"* welcome sign, then wind through columns of tourists—paying our entrance fees and declaring that we have brought nothing to harm the creatures we will visit.

Though the sky is overcast, the air is hot and arid. Like a synchronized chorus line, we shed jackets and overshirts and

The Dream

stuff them into the pockets of our backpacks. I feel a giant "hurrah" building inside me, but swallow it. I want to sing like Barbra Streisand, jump like a two-year old, run beyond the airport crowds across the sand to the sea. But I manage to restrain myself.

"My name is Francisco," a voice bellows through the surge of passengers. About five-feet-seven with glistening tan skin and a stocky build, he musters attention: "I am the guide for the yacht *Beluga*. Bring your luggage here, and wait over there for the bus that will take us to our ferry. We will be crossing the Itabaca Channel to the highlands of Santa Cruz. Then we'll head into the city of Puerto Ayora and travel by dinghy to our boat."

9
Hearts in the Highlands

Friday, August 18
Late Morning in the Highlands of Santa Cruz, a Central Island

As our bus speeds along, I surrender myself to volcanic mountains, a turquoise sea, and land dotted with cacti. A short trip across the waters to Santa Cruz and we board another bus. My mind in a dream state catches only bits and pieces of Francisco's perfect English narration. I slide my window open and inhale the moist air. Less than an hour ago, I peeled off my jacket in the dry heat of Baltra. But here in the highlands, the *garúa* mist that envelops the island in the cool season is all around us, bringing with it a damp, gray stillness.

"Ecuador," he proudly informs us, "declared ninety-seven percent of the Galápagos as a national park in 1959. It has been a major tourist attraction since 1970. Santa Cruz is one of few islands where all seven vegetation zones are represented. We will de-board now and walk through the *Scalesia* rainforest zone."

While the island may look green now, he tells us, the vegetation is dry and brown compared to what it will be in their spring and summer. I absorb the quiet—interrupted only by bird calls as we weave our way through grasses and giant ferns. Everywhere we look are eerie *Scalesia* or giant daisy trees from the sunflower family. "There are at least fifteen species of this plant on these islands," Francisco tells us. "They have evolved from one species and grow nowhere else in the world." At least thirty feet tall, these rare trees struggle for dominance among foreign species like the Cuban cedar, red quinine, guava, and passion fruit—all brought here by humans. Hiking the muddy terrain through a haze of greenness, we gaze upwards at the graceful, sinuous branches draped in silvery moss.

"Watch your step." Franciso's voice issues an emphatic warning. Seconds later we arrive at the rim of a pit crater more than four hundred feet deep, one of two giant sink holes known as *Los Gemelos* or the twins. I stare down into the huge hollow and remember the crater village. Only this time, instead of tiny parcels of farmland, a gigantic amphitheater of vegetation teems with life, wild and verdant. I move away from the crater, imagining how it came to be. "No eruption took place here," Francisco tells us. "A burning column of molten rock called magma rose from within the earth and penetrated the earth's crust, forming a lake of bubbling molten lava. Over time, the ground collapsed, leaving this giant crater."

Cruising along on the bus, I look out on a blurry backdrop of yellow warblers flying among the trees. Seconds later another primary color bursts onto the scene. High on a *Scalesia* branch, a male vermilion flycatcher perches—his vibrant red breast and head, black mask and wings commanding attention. The bus pauses while Francisco points to a small, brownish bird roosting in a tree branch about twelve feet off the ground. "Check him out. He may not look impressive, but he is the famous woodpecker finch, one of Darwin's thirteen finch species."

Hearts in the Highlands

My heart races. There sits the tool-making wizard of a bird that started me on this quest. If only he'd grasp a twig and dig for grubs like the finch on the television special. But he just sits there surveying his world—tool making definitely not on his agenda at the moment. I watch him intently, this ordinary looking brownish creature with his short tail and puffed, slightly speckled belly. But he isn't ordinary, and I know it. He's a feathery descendent of the wild ones that helped the great scientist understand how animals and plants split into different species.

We continue our tour, stopping to observe a vegetarian finch examining bark that he may later strip and curl like the shavings on a wood shop floor. He will not actually use the bark, we learn, but peel it away to reach the sugary inner layers of the stem. Unlike the *Scalesia* and foreign invader plants we saw just minutes ago, these two finches don't compete with each other. One eats insects and grubs while the other feeds on buds, leaves, fruits, and blossoms.

Each has its own niche while the *Scalesia* must struggle each day for its share of sunlight and nutrients. I've never before thought of conflict among plants. Yet here I am in a kind of war zone where battles between species take on gigantic importance, and the demise of any one plant or creature can devastate the fragile balance. If invaders succeed in choking out these daisy trees, what creatures that depend upon them will be affected? And will they evolve or simply disappear?

Out of the bus again, we listen to Francisco's first lesson on the origin of species in the Galápagos where life has either flown in, blown in, floated in, or been transported by humans. "Native," Francisco emphasizes—like the pelican and the yellow warbler—live here, but are common to other

Portrait of a Galápagos Dome-shaped Tortoise

places as well. "Introduced"—like the destructive goats, pigs, dogs, and rats, or the luscious guayabas and passion fruit—have been brought to the islands by humans, some of whom like Francisco himself, have chosen to settle here. "Endemic" means that an organism, like the woodpecker finch or the Galápagos tortoise we can see just ahead of us, exists here but nowhere else in the world.

    I stop breathing for an instant. Nothing has prepared me for my first sighting of dome-shaped tortoises. I lose track of Francisco's narration and watch the massive creatures, like armored vehicles, advance through the highland grasses. Beneath their arched shells—their huge shoulders, legs and clumplike feet carry them forward on a food-finding mission. A few stop along the way to graze, pulling blades of grass, roots and all, out of the soil. "On other islands which you will unfortunately not be visiting," Francisco says, "some of the tortoises have shells shaped like saddles. Saddleback tortoises are not as large as these, but their necks and legs are longer."

Standing among the domed ones, I regard their slanted eyes and weathered skin as the grass bends and swishes beneath their elephantine feet. Some say you can estimate a young tortoise's age ("young" meaning ten years or less) by counting the growth rings on its shell, but I see rings far too close together for counting.

We break off into groups following single tortoises along their paths, but the quiet giants pay no more attention to us than they do to each other. After a few minutes, Francisco calls us together. "Their ancestors," he says, "may have floated here from the South American mainland on storm-tossed rafts of vegetation and hunks of earth. They might even have floated here on their own. Unlike turtles, tortoises are not fine swimmers, but they do love to wade and could have been carried to the islands on ocean currents."

I recall how in his *Beagle* diary Darwin marveled at how fast tortoises can walk: He wrote how one large one "walked at the rate of sixty yards in ten minutes, that is 360 yards in the hour, or four miles a day,—allowing a little time for it to eat on the road." They walk their fastest, we learn, when they're migrating up or down the island during nesting season.

These dome-shaped tortoises live here in the highlands, Francisco explains, but lay their eggs in the lowlands which have ideal soil and because incubation there tends to be more successful. Some experts believe that eggs laid during the coolest part of the cool season produce males while those laid as the climate shifts to the hot season produce females. When I ask Francisco about this, he explains that the tortoise has no sex gene. The temperature of her nest determines the

Bette with Galaápagos Dome-shaped Tortoise

sex of the hatchlings. If a nest has a blend of temperatures, both male and female offspring will probably result.

I crouch to view a passing giant, eye-to-eye, but it meets my inquiring glances with disinterest. How it has learned the secrets of survival is a mystery to me. Certainly not from its mother, I know. A baby tortoise must fend for itself, for it has no role model. Like most reptiles, its mother abandons her eggs, leaving incubation to the sun and rearing to nature.

Back on the bus again, I smile at the "Peligro Cruce de Tortugas" sign—as rudimentary here as "Caution: Deer Crossing" throughout the U.S. The bus bounces down a narrow road bordered by a two hundred year old cedar and prolific, introduced avocado trees. A light island rain sprinkles my face—and the ever-changing weather oscillates from warm to cool. We stop again and de-board close to snowy white cattle egrets hunched over and resting on the branches of a reddish flame tree. "We are now in the Transition Zone," Francisco announces. Amidst water-fern greenery, we listen to the hissing sound of tortoises, once close to extinction. Though tortoises don't actually hiss (they don't have diaphragms as we do), they make a sibilant sound as they expel the air in their lungs by drawing in their legs.

"Be careful when you approach them," Francisco warns. "That sound you hear shows they are aware of and perhaps nervous about our presence." While this is a reserve for the tortoises, I know they do not have a safe harbor here. The park must send wardens to the nesting zone to kill pigs and other predators that threaten their eggs and their young and to protect them from poachers.

"Poaching is a serious problem here on the islands," Francisco tells us. "Poor fishermen who want a longer fishing season, better roads, or higher catch limits threaten our tortoises. They hold them hostages or kill them as a political ploy because they know how important the tortoise is to tourism." Though these tortoises appear safe, I wonder what our presence brings that could ultimately put them in danger. A virus? Bacteria? Insects hidden in our clothing?

Or a physical proximity that may make them unduly secure around our species.

"You'll notice," Francisco points out, "that the few flowers you see are mostly white and yellow. In the Galápagos, we have no hummingbirds and very few pollinating insects—mostly carpenter bees and some butterflies." Without pollinators, large, colorful flowers don't tend to grow wild. Francisco's simple observation intrigues me. By attracting pollinators, colorful flowers increase their numbers and their potential for survival. Without pollinators, there's no advantage in expending the energy needed to produce colorful flowers. I recall Darwin's remark: "All the plants have a wretched, weedy appearance, and I did not see one beautiful flower."

Pushing our way through tree ferns with curved branches and plastic-like fronds, we stop briefly to admire a teak tree native to Southeast Asia. Though beautiful, it may endanger both endemic and native vegetation. Tired from our walk, we stop in a thatched, protected area. Two women from the village of Puerto Ayora have set out small china cups of lemon grass tea on cloth-covered tables. How rare to be here in the highlands a few feet from sibilant tortoises—sipping tea. The irony of it makes me smile.

The quiet here reminds me of my favorite Yeats' poem: *And I shall have some peace there, for peace comes dropping slow, dropping from the veils of the morning to where the cricket sings.*

I want to speak with Suzanne about what we're experiencing, but find her deep in thought. She stares at the tortoises, then writes in her notebook. Watching the muscles in her face tighten, I sense a tempest looming.

10
Sights of the City

Friday, August 18
Early Afternoon in Puerto Ayora, Santa Cruz

Our bus pulls us up a steep grade to a muddy slope and then onto the main road. Wind-blown rain spatters the windows and speckles our view of a white-cheeked pintail duck winging its way to the coast. Once an undersea volcano, the island undulates with civilization: huts shaded by banana and papaya trees heavy with fruit; orchards of avocado, grapefruit, and oranges. Goats, dogs, pigs, rats and cats (though I see none) are here now—introduced by early settlers. Through a low cloud cover, we see expanses of bamboo, grass, and broad-leafed plants like the ones that grow wild under our back porch at home. My socks are wet, and my shoes—which eight-year-old Oscar polished beneath the giant Virgin's statue in Quito—are caked with mud.

I listen as my teacher-naturalist daughter, Suzanne, banters with Francisco. "We were too close to the tortoises," she tells him. "It's not good for them."

"And how would you know?" he chides.

"They were stressed. You yourself told us the sounds they made could be because of nervousness."

"Trust me. I care what happens to them, and I know what I'm doing. Look over there. We are passing close to my house. We're right on the fringes of Puerto Ayora."

My daughter stops challenging Francisco for the moment, but I know he hasn't heard the last from her. I understand her concern and admire her courage to speak out. Even Francisco, experienced and committed to conservation as he is, might sometimes get caught up in the entertainment aspect of his job.

A helmeted motorcycle driver streaks past our bus. Small cinderblock homes with flower gardens and lines of billowing laundry edge our path, along with abandoned stone dwellings, a school with bleachers, and a soccer field. Minutes later, we're in Puerto Ayora, with its gasoline station, cobblestone streets, children with pet goats, a burger stand, a supermarket, hotels, and several Internet cafes. The bus drops us off at the dock where Francisco tells us we should meet him in about an hour.

Now we have time to explore the town. The four of us walk the sunny cobblestones and enter a miniature city of standing white gravestones. Dry leaves crunch beneath our shoes as we walk among dozens of crosses and read inscriptions carved or hand-painted on monument walls. Altars to the lost abound with statues, photos, and vases of colorful blooms that never require water. Occasional wild flowers sprout in earth-filled reservoirs bordering the graves.

We stop before the graves of Miguel Suarez who died in 1968 at fifty-two and Roberta L. Rivedeneira Peña who lived only eleven years. A purple hydrangea wilts beneath her name. To its side, adjacent to the year of her death, someone has tacked a note, now faded and unintelligible, but still evoking images of her short life. The idea of a child dying and

Puerto Ayora Cemetery

parents living on and perhaps regretting what they did or didn't do enters and leaves my mind quickly—too terrible to contemplate. In this intensely personal cemetery, the dead seem to interlace the living.

Across from the cemetery, we browse tiny roadside stands where we buy postcards and a few blue-footed booby magnets, but refrain from buying T-shirts despite pleas from smiling women in starched white blouses and colorful skirts. We walk a few steps and enter the Charles Darwin Research Station to visit[1] Lonesome George, possibly the only surviving pure blooded Pinta Island tortoise. George, possibly sixty to ninety years old, is a solemn reminder of the

---

[1] In 2008, fourteen fertile eggs were discovered in George's corral which he shares with two closely related female tortoises from Isabela Island. The possibility that George's DNA could be passed to a future generation of tortoises caused great excitement, but the eggs turned out to be infertile. Despite this unfortunate finding, efforts to repopulate Pinta Island continue. Go to the Galapagos Conservancy Web site (www.galapagos.org) for current information.

destruction wreaked by whalers and sealers. I watch as he stares in our direction and then slowly turns his almost two hundred pound body. On the brass plaque before his cage, the sign tells of researchers who futilely tempt him with mates. Despite their many efforts, he seems to run out of steam around female tortoises closest to his own sub-species. Since he could live another hundred years or more, there is time for him to choose a mate.

Walking away from George, I think about what humans have done to cause his plight and what the Charles Darwin Foundation and the Galápagos National Park are doing to protect him and all the other species on these remarkable islands. I make a mental note to start supporting Galápagos conservation efforts when I get home, then move on to see young tortoises from both saddlebacked and domed populations. Although dependent on humans for their survival now, they will be returned to their islands of origin when they turn four years old. There they will grow and live wild like the ones we saw today grazing in the highlands.

Heather and I leave the research center and meander along the road. "Mom, did you hear Suz on the bus with Francisco?"

"He seemed to handle it pretty well, but he did look a bit surprised," I laugh.

"I just hope she doesn't overdo it. You know how she can get."

Minutes later Suzanne and Yolis run toward us breathlessly talking about what I've been thinking—the shock of seeing Internet cafes, ATMs, discos and taxis here in Darwin's "mystery of mysteries." I've seen pictures of Puerto Ayora and read about Santa Cruz's growing population, but reading about a place and being here are entirely different.

Sights of the City

Puerto Ayora Vendor

"We were asking one of the shop clerks about the beautiful tortoise sculptures," Yolis tells us, "and he started talking about how the poor people here depend on tourist money and money from fishing and how the park makes rules like fishing-limits that hurt the people. He says the Ecuadorian government cares more for animals and money than for people."

I think about the T-shirts I didn't buy and the women at the vending stall. At the time, all I was thinking about was my tight budget, but I could have sprung for more than I bought. Competition is everywhere. Animals vs. animals; plants vs. plants; people vs. animals and plants *and* the government. The government needs to protect its citizens, but also must protect the creatures that bring tourist dollars. The wildlife needs government protection *and* respect from the residents. The people need government protection so they can earn their livelihood, but they also need protection for the park so that tourists keep coming. And if people feel less important than animals, they may retaliate by poaching and taking hostages.

Damn, I should have bought T-shirts, but it's too late now. Our time is up.

11
Seaward Bound

Friday, August 18
Afternoon and Evening—Drifting at Sea

Francisco is waiting for us on the dock of Academy Bay. As strangers in this land, we find ourselves transfixed by a promenade of life that residents consider routine. Strolling marine iguanas, oblivious to our presence, absorb the equatorial heat while flamboyant purple-red crabs scavenge for tidbits on the black lava rocks. Brown pelicans, staring seaward, stand motionless—poised for the hunt. And in the air beyond us, dive-bombing booby birds shoot into the sea. We are shocked into reality when Francisco directs us with military precision into two flat-bottomed, motorized dinghies or *pangas* like the ones my teacher-friend Jim described so many years ago.

"When you get aboard, put on your life jacket," Francisco tells us. "You must never be on board without it. Take my hand with your left hand—and move."

I have no time to think, no time for questions. I step off the stable wooden platform, take his hand, and hop down into the gently rocking rubber raft—immediately sitting, sliding over, and grabbing the life jacket passed to me. Others follow, each sighing when securely in place. I glance at my misty-eyed, smiling girls and cousin.

Boarding the Beluga

"No more than four to a side," he thunders when one of our group appears uncertain. "We must keep our balance." When a man ignores his directions and bounces into the dinghy precariously without taking his hand, he blares: "I said, 'Take my hand!'"

With eight of us aboard, we motor off, but without Francisco. As we pull away, I watch him at the dock stretching out his hand in staccato movements over and over. He springs aboard behind the driver and soon our fellow travelers are gliding toward us across the easy sea. The ocean spray seeps into my pores and the still-foreign smell of salt begins its slow mutation to familiar.

Five minutes later we are at her side—the *Beluga*—our three-deck, shining white home for the next five days. Our companions in dinghy number two arrive seconds later, and Francisco leaps from them to us, steadies our dinghy and secures it to the boat. Members of the crew are on deck to welcome and help us. When Francisco is ready for us to disembark and climb the metal stairs to the one-hundred-ten-foot ship, I move rapidly, with no show of fear. I do so because he expects no less and because if I falter, I will break the mandatory rhythm of debarkation. Again, I take his hand—jump from the raft to the steps, pulling myself up till safe on the deck.

"Take off your life jackets and pass them to Reynaldo. Take your shoes off, shake them, and leave them in the bins at the top of the ladder," he tells us. "We don't want sand in the ship."

Aboard the *Beluga*, a crew member directs us through the gray-blue dining room to our cabins at the rear of the main deck. Our tiny stateroom, at first sight, appears bigger than it is. Two large, adjoining picture windows facing the stern and starboard sides seem to extend our living space beyond the ship into a turquoise sea. In actuality, we probably have less than seventy square feet, but more than enough room for our backpacks and belongings. Our cabin consists of a

Yolanda in Stateroom

narrow closet, two walnut bunk beds against one wall, and a tiny bathroom. Heather and Suzanne peek in to check out our surroundings and then return to their cabin to unpack and rest. While Yolis cleans up, I kneel on the carpet and touch my nose to the glass—like the child in the Hudson Sedan. A brown pelican has hitched a ride on the railing—probably waiting for bits of our dinner, its cooking aroma just now permeating our space.

At seven, we dine with our fellow passengers. Yolis and I sit close to Pearl, an educator from the Bronx and her husband, Henry, a judge. Across from us are Jennifer, a hospital administrator from Oregon, her E.R. doctor-husband, Bill, and their fourteen-year-old David who will begin high school in a few weeks. At the other table, Suzanne and Heather join Deborah, an Indiana internist; her biologist husband,

Matthew, and their twelve-year-old Monica—along with retired Heidelberg chemist Steve; his Swedish wife, Sandra; and five-year old Cathy.

"My name is Alfonso," our waiter announces. "I am here to serve you." He is tall with dark curly hair—no more than twenty-five. His tan skin glows in the cabin light as he fills our glasses with water and answers our questions about the food, the weather, and his tenure on the boat. "I have worked here for a year," he tells us. "I like it very much. " All but David, who has a painful sore throat, take multiple servings of rice, vegetables, and *bacalao*, a wonderfully mild codfish. In between bites, we talk about whether David should be taking French or Spanish in high school, about the fine restaurants in Quito, and about the unrealistic law and medical dramas on TV.

I am impressed, not so much with our topics nor with the details of our conversation, but with the ease of our words and the closeness and feelings that pass among us. We are, after all, drawn to these islands by a common love and respect. Francisco stops at our table to tell us that we must arise "no later than six-thirty" and be at breakfast "no later than seven." The only meal that will be at all flexible is lunch—which will usually be between one and two—depending on when we come back from our morning island.

After dinner, we are the only ones to take Reynaldo and fellow crew member Gabriel's invitation to go back to Puerto Ayora. The two of them plan a bit of partying while we want to take this last opportunity to e-mail home and see the town at night. After skimming the waters from the *Beluga* to the dock, we wave goodbye to our crew friends and promise to meet them in an hour and a half.

Finding an Internet cafe is a simple task in a town that thrives on tourists. After an agonizingly slow e-mail connection, we drink beer in a *taberna* and polish off a bag of cookies bought from a street vendor. I see no T-shirts for sale, but buy several more post cards to support the local economy. Yolis and Suzanne set off to browse among the cafes while Heather and I walk the glittering shoreline. Midway during our walk, we enter a naval base by mistake and sprint away when a military guard approaches us to check our credentials and discover our business. *Nada, nada, Señor—Perdónanos*, "Nothing, nothing—Pardon us," we call to him. From street lamp to street lamp, we run through Waterfront Park to the main dock where we meet up with Yolis and Suzanne.

Ten minutes later, our handsome Ecuadorian companions return to ferry us to the ship for our first night. On the black sea, under the equatorial moon, Reynaldo and Gabriel tell us about their *"tiempo grandioso"* drinking, singing, and dancing, and we tell them about our run-in with the military police. "Sing for us," we beg them, but they shyly refuse. *Otro tiempo, Señoras*, they laugh. "Another time." The breeze is warm and the sea calm, and we share stories with one another in partial, broken, and perfect Spanish and English until our raft touches the *Beluga* and we are home.

At about ten we pass through the ship's empty dining-lounge area to reach our bedrooms—the only ones on the main deck. In our dinky bathroom, I wedge myself between the sink and toilet to brush my teeth and stare into the mirror at the deep circles under my eyes. After all these dreaming years, I still can't believe I'm here—inside the pages of *National Geographic* and *Smithsonian*. Gentle though they are, the waves buffet me backwards and forwards, and I must steady myself by clinging to the sink edge. The door

Seaward Bound

swings in and out banging against the wall, and I grab its knob and pull it toward me. By the time I stumble back into our room, Yolis is quiet in the top bunk. After an initial hangover sensation, I climb into my bed, an arm's-length from the floor. I wake only once to the sound of a jangling chain when we pull up anchor during the night.

12

A Red Beach

Saturday, August 19
Morning on Rábida, a Central Island

Uncharacteristically, I awaken to the buzz of my watch alarm even before Francisco's six-thirty knock. Listening to Yolis' rhythmic breathing, I roll out onto the carpet and shuffle on my knees to the window to watch the sun reflected in a rippling sea. A golden mist hovers over the steel-blue waters. How lucky we are to have windows while others must peek through portholes or go out on deck. I see land about five dinghy-minutes away and realize that we have reached our first destination.

After breakfast, Francisco summarizes the day's itinerary: Rábida Island in the morning and Santiago in the afternoon—with time for snorkeling in between the two. I am nervous about the snorkeling, but ready for the challenge with my disposable contact lenses and the wet suit rented in Quito. "Be careful to wipe your shoes clean in between islands," he cautions. "We must prevent the transfer of seeds and soil from one island to another whenever possible."

We make a "wet landing" on Rábida, sometimes called the geographic center of the Galápagos. Wet landings require us to wade a few feet from our *panga* to the shore. Stepping barefoot onto the sand, we put on our shoes and walk along the iron-red beach, careful not to disturb the residents. Maternal sea lions bark, their little ones nursing and bleating like goats. Five-hundred-pound beach master males with raised bumps on their heads bellow and roar, warning other bulls to stay away. These masters stake claim to the best territories—parcels of beach that attract the most females. They defend their boundaries aggressively against competitors, and when attack by white-tip reef sharks and hammerheads is imminent, they often guide defenseless pups to safety.

A strip of mangrove trees separates the beach from a deserted, brackish lagoon—once a paradise for flamingoes. Occasionally, flamingoes still visit this lagoon but their food supply of algae and crustaceans has dwindled—probably because of El Niño or possibly because sea lions like the noisy ones we're watching have polluted their waters. We watch a blustery bull wander away from the herd on the beach to slosh alone in the salty pond. Why he chooses a solitary escape is something only he understands.

We stay on the beach, watching the playful antics of sea lion pups and their mothers. We are respectful of their territory, doing nothing to make them nervous or defensive. I pause to appreciate a young one's fur—brushed golden red-brown with an almost turquoise hue. They may touch *us*, Francisco emphasizes, but we must *never* touch them.

Along the way, we pass an abandoned pup with a bleeding front flipper probably torn by a hammerhead. What was his condition before the attack, I wonder. Was he slower or weaker than the others in the nursery, leaving the more physically fit to create offspring like themselves? Or was he simply in the wrong place at the wrong time? As never before, I ponder the mystery of survival. I doubt that Darwin would

A Red Beach

Mother with Galápagos Seal Lion Pup

have dwelt on the survival of a single pup. He would have been concerned about the entire sea lion species and the special traits that have made them survive through the ages. He would have analyzed their powerful swimming skills, their uniquely finned feet, and other traits that give them an edge over competitors. But Darwin was a scientist, and I'm not. I can't stop personalizing the fate of this one, poor creature. Walking slowly past him, I am careful to bring no more distress. I accept my powerlessness. I have no dominion here.

Minutes later I hear Suzanne's raised voice. She's talking with Francisco and her tone is emotional. "I don't understand how you can let us get so close to the wildlife. We walked right past that injured pup. It was bad enough that he was hungry and probably in pain, but on top of that he had to endure a horde of tourists walking through his territory."

"Just you wait," Francisco shoots back at her. "You paid good money to come here because you care about these animals. Are you naïve enough to believe that the government would have declared ninety-seven percent of these islands as a park without a payoff? You are a purist, and I admire that, but you're going to have to let me do my job. These creatures will never be placed in jeopardy while I'm in charge."

Suzanne and Francisco walk side by side, still talking—the volume of their voices fading to inaudibility. The three of us strain to hear, but give up.

"Do you think she'll ever let up?" Yolis asks and puts her arm around me.

Heather shakes her head. "Not when she's passionate about something. This isn't the first or the last time my sister will embarrass me."

I feel stress building and though I want to make everything fine, I have no more control now than I did with the sea lion pup. Minutes pass and the once contentious Suzanne and Francisco begin laughing.

"Looks like the challengers have discovered common ground," I announce.

We wind our way up a red lava rock path lined with four- to twelve-foot cacti, *palo santo* trees, and salt bush. Yellow warblers, the exact color of the salt bush, flit from plant to plant. Looking down a cliff to a small ocean inlet, I inhale the sea air and watch the sally lightfoot crabs. They move so rapidly that their tangerine-colored shells splotched with gold and magenta almost blur against the gray-black rocks. I remember John Steinbeck's observation about these agile creatures in *The Log from the Sea of Cortez*: "They seem to be able to run in all four directions."

Along the cliff sides, American oystercatchers pry open delicacies with long, red, spiky bills, and Galápagos mockingbirds bounce and chase atop the rocks in search of others'

A Red Beach

American Oyster Catcher

food. I haven't studied the springy movements of mocking birds since college when my parents rescued an injured one from a hungry tabby cat. I can still picture the rehabilitated "Waldo" swooping onto the breakfast table and absconding with smidgeons of dripping egg yolk or his beloved buttered toast. He roomed with us for about a year until the day when my mother clothes-pinned his cage to the laundry line and a gust of wind knocked it to the ground, opening his door to the life mockingbirds are meant to live.

"During a drought," Francisco tells us, "these curious wonders survive while others starve." Is it any wonder, I think to myself.

Minutes later back in our *panga* among my companions, I reflect on what I've seen this morning. Unlike visitors to the San Diego Wild Animal Park, we have not been protected from unpleasantness. Here, injured and dead creatures are not whisked away before we come onto the scene, and no human can intervene on their behalf. Although I prefer to envision healthy sea lion pups and block out the scene of the abandoned one, I am beginning to accept both realities. No film or journal article could have prepared me for this.

13
Snorkeling for the Meek

Saturday, August 19
Late Morning in the Waters off Rábida

"If you're going snorkeling, come up on deck now," calls Francisco. Grabbing my rented wetsuit, I follow the bare feet and long, golden legs of my girls up the almost vertical, ladder-like stairs. My history of swimming in heated pools with lifeguards ready to scoop me from danger has not equipped me for this. Nor has my playful wave jumping—a few feet off the La Jolla shore. I stand close to the rail as Francisco and Alfonso distribute the gear—assembly-line style. Fins, masks, and wetsuits glide from person to person. "We've already rented our suits," Yolis tells them. "All we need are masks and fins." Francisco passes me a mask, but can't find a pair of fins to fit me. "Don't worry," he says. "For beginners, it's not a problem. You'll stay close to the shore and can snorkel without them." To prevent our masks from clouding, he directs us to swab the insides with saliva or toothpaste and then wash them in the sea.

Along with the others, I wriggle awkwardly into my wetsuit, and then just as awkwardly practice putting on my mask. Alfonso stands by as I pull the mask strap down toward the nape of my neck and then try to maneuver the goggles up to the right position. He gently presses the goggles forward against my face to test for water-tightness. *No, Señora, necesitamos moverlos.* "We need to move them." He repositions them, pushing in again—suctioning my eyes, and adjusting and readjusting the strap. *OK, Señora, es su turno.* "It's your turn." We laugh at my clumsiness, and just when I am catching on, I hear Francisco's proclamation: "We will leave in exactly ten minutes!" No time to spare—I quickly take off the mask and dash to my cabin to put on newly purchased disposable contacts.

I have recently switched from hard lenses to regular glasses, but for this trip, my optometrist has convinced me to try the disposables for snorkeling because they're "so much more convenient." Minutes later, in the bathroom, Heather, a veteran soft lens wearer, helps me insert my right lens. I try to insert the left by myself, but after seconds of nervous frustration and nausea over touching my eyeball, I decide that seeing with one eye will have to do. Just in time for de-boarding, I join the others and we dinghy out to the area where we will snorkel. I dip my fingers into the chilly waters and think of Pearl who has forgone this adventure. "You're very brave," she told me earlier during a conversation about her trips to Cuba and the Amazon. "I snorkel only when it's warm." Somehow I do not feel brave.

Francisco goes with the more experienced divers who need no protective shore nearby. Though an experienced snorkeler, Yolis stays with me—close to the shore. Entering the water, I feel the numbing, seventy degree cold on my feet, legs, knees, and uncovered arms. I put my face down, open my eyes, and float outward, kicking my bare feet. As instructed by Yolis, I breathe through my mouth, emitting hollow, echoing, animal sounds that connect me with this new world. All seems easy at first. Through the clear waters swim schools of small grey Creoles and angels—black with white vertical stripes and orange-yellow tails. About an arm's length away glide clouds of blue parrotfish and yellow-tailed surgeonfish like the ones I've admired many times in aquariums. Dazzled and slightly tipsy, I want to share my excitement with someone, but must breathe through my mouth in semi-silence. I groan aloud just to hear the sound of a human voice I can exclaim to, not minding that the voice I hear is my own.

"How are you doing?" my cousin asks as I finally raise my head above water. And I tell her what I've seen. Minutes later, when I re-enter the water world, my breathing pattern has been disrupted, and I can't recapture my rhythm. My mask clouds in spite of spit, and I take in briny liquid. Raising my head again, I take stock, calm myself, and straighten my snorkeling tube which is twisted and pointed downward like a straw for sucking the sea.

I re-enter the water, peeking into the private caves and crevices of green sea urchins, purple sea stars, and golden sea fans. Suddenly the colors stop and I can see no more creatures. Instead, loose grains of disintegrated lava permeate the water like an underwater sandstorm. As the liquid haze envelops me, I struggle to get beyond it to the light. I'm too far from shore. I must go back. But shadowy black canyon-like ledges loom and the gray, frigid waters have reached depths I'm not prepared to travel. A rock protrudes from within the fog and I stretch my fingers toward it. I push off against its jagged surface into a strong current. Like a tiny handball flung against a wall, my body rebounds. My heart beats erratically and I push harder against the rock, this time sending myself pulsating into safe waters. A yellow-tailed damselfish swims by, unaware of me and my trembling. The colors resume and I am once again in the company of reef dwellers. I respect the current. The sea is not my home. It is theirs.

Resurfacing, I hear Yolis' laughter. "A sea lion just swam through your legs," she tells me. And to think I missed it. The more adventurous swimmers, back from deeper waters, join us in our cove. Heather calls out that some, including Suzanne, have seen rays and turtles and a whitetip reef shark. I am thrilled for them but ecstatic over my own victory, however small in comparison. Bone cold—I am happy to climb onto rocks where hermit crabs carry their own shelters. Entering the dinghy minutes later, I rub the rock indentations on my palm and feel numbly warm and protected.

14
God's Creatures and Lava

Saturday, August 19
Afternoon in Santiago, a Central Island

Back aboard the *Beluga*, we shower, dress, and eat. Pearl and I spend time together on the top deck watching two bottlenose dolphins race through the water alongside the ship. Our hands on the railing, we lean forward when a dolphin we hadn't noticed splashes to the surface. We talk of the sea, our children, and the whole Galápagos experience. "You were daring," she tells me, "going into that frigid water." "You're the brave one," I tell her, "hiking through the insect-ridden Amazon jungle."

The more we're together, the more we discover those elusive qualities that forge friendships. She talks about her son, a mystery writer who's published several books and is writing another. "What about your girls?" she asks. "I see flashes of their mother in each of them."

When I ask what she means, she takes my hand. "Suzanne's passion for righting wrongs, of course. I could see you were embarrassed when she took on Francisco, but you almost nodded at what she was saying and how she spoke out. And Heather's sensitivity to you and everything around her. She ran to help you with your contact lenses because she didn't want you to miss out on anything. Everyone's excited to be here, but with you the excitement seems deeper. "

Pearl's dark eyes absorb more than words as she listens to my dreams about this trip. "It's almost like you're home on these islands," she says. "I sensed that about you almost from the first moment." We laugh at all the experiences we've shared. She supervises student teachers; I'm a former teacher and an education writer. She's a New York club jazz singer; I sing with a group that performs in retirement and nursing homes. A seasoned adventurer, Pearl has stories to tell. A novice traveler, I am eager for stories of my own.

In the afternoon, we slosh through another wet landing—this time on Santiago where Darwin and his small party stayed a week "whilst the *Beagle* went for water." Darwin walked this very island a hundred and sixty-five years ago. Galápagos was in summer then, and the temperature of the sand had reached one hundred thirty-seven. Possibly even hotter, Darwin wrote in his journal, but the crew's thermometer wasn't "graduated any higher" and so they couldn't be sure.

Our group walks along the sand, barren and black, but cool compared to the sand Charles Darwin trekked. Amidst squealing American oystercatchers, we look out onto steep-sided lava cliffs that appear layered with petrified wood. Dozens of brown pelicans—some in flight, others perching on craggy bluffs—scan the waters for fish. Paces away, adult sea lions roll in the sand and bark while a bleating baby pup, umbilical cord still dangling, waddles alongside its protective mother.

"In about a week, the umbilical cord will fall off," Francisco tells us. "We must avoid clapping sounds," he stresses, as we pass among them. "While they may not fear us, they do fear the loud noises we humans can create." He recounts a story told by guides on the islands about how long ago murderous men clapped sticks on their hands to show power, and how fear of this sound still haunts these innocent beings. "A charming story that helps protect the sea lions," whispers Suzanne, "but I doubt it's true."

Hiking through the barren lava, we hear stories about feral pigs and goats, but see none. For many years, pigs decimated sea turtle and tortoise nestings on Santiago. They dug up their nests for eggs and also stole eggs from ground-nesting birds. Goats have destroyed vegetation here—turning shrubs into sticks and uprooting protective grasses, sometimes causing massive erosion. "Goats are still here—hiding," Francisco tells us. "Make no mistake. They are watching you from way out there." He gestures toward a bleak infinity, and as we advance along the trail, he parcels out a tale of intruders in the late 1800s or early 1900s bringing goats to this island. "They multiplied fast until they numbered about seventy thousand."[1]

We pass yellow-crowned night herons, Galápagos doves, a soaring Galápagos hawk, and a salt mine operation of old. On Santiago, Darwin spoke of a "salina" or "lake from which salt is produced." Accompanied by "a party of Spaniards" he walked "over a rugged field of recent lava" to a tuff-crater enclosing the salt lake. There in the bushes they discovered the skull of a poor captain murdered by his sailors and left here in "this quiet spot." Lingering before this reminder of Darwin's voyage, I see no skull but feel a chill at being so close to where he once passed.

---

[1] In May 2002, Santiago was declared "pig free" after an extensive eradication by the Galápagos National Park Service and the Charles Darwin Foundation. By 2005, the island was also declared "goat free." While Judas goats (sterile, radio-collared goats) are still on Santiago for monitoring purposes, the island no longer harbors feral goats.

Galápagos Marine Iguanas

"Keep a respectful distance from the marine iguanas," Francisco calls out with a backwards smile to Suzanne. Just minutes from the salt mine, we now overlook a teeming sea and must maneuver across rocks crowded with leathery black marine iguanas sneezing excess salt through a gland connected to their nostrils. Stacked one atop the other for warmth and for other possible advantages not yet understood, they resemble a lava rock formation with scaled outcroppings—the males with their spine crests, males and females alike splattered with salt crystals. At night the iguanas continue their stacking behavior to avoid ticks and perhaps other parasites. A lone iguana, we learn, gets more ticks.

"They are strong swimmers, and the larger adults can dive to depths of more than thirty feet as they seek out red or green algae-type seaweed," Francisco announces. "When they arrived here millions of years ago, these iguanas may have been green. Some scientists think they adapted to black so they could retain more heat."

I watch ocean waters rushing beneath powerful claws that cling for stability to the rocks and recall Darwin's observation that when frightened these creatures do not voluntarily escape into the water. Initially, he referred to the iguana's stupidity: "Hence it is easy to drive these lizards down to any little point overhanging the sea, where they will sooner allow a person to catch hold of their tails than jump into the water." Later he said that their only natural enemies are sharks: "Hence, probably, urged by a fixed and hereditary instinct that the shore is its place of safety, whatever the emergency may be, it there takes refuge." I ask Francisco about Darwin's assumption, and he tells me that some scientists think the iguana Darwin saw might have stayed on the rocks for warmth. Having just left the water, it might have been too cold to re-enter.

"There are no true seals in the Galápagos," Francisco calls out as we approach the fur seal grotto. The biological term "true seals," we discover, refers to a specific family of pinnipeds or fin-footed creatures that have ear holes. Francisco doesn't mean that *real* seals don't exist in the Galápagos. The Galápagos fur seals we're watching now have external ear flaps and are from a different family that includes the sea lions we saw this morning. Looking down into two sheltered pools connected by a black lava bridge, we see large-eyed creatures splashing and gliding—their front flippers acting like paddles. I hear no barking sounds as I did on Rábida. Instead, these thick-furred brown beings seem to cough and groan to each other. Several females bulky with young will be tending their pups soon. Both the fur seal and sea lion, though not technically endangered, are being monitored closely by the Charles Darwin Foundation. Illegal fishing practices as well as environmental factors like El Niño keep them in jeopardy.

Balancing on the weathered rocks, we overlook an aquamarine "pothole" called "Darwin's Toilet" because it empties and fills with a familiar swishing sound. A male fur seal crawls out of the bowl onto the lava to rest. We too are ready for rest.

A spiraling line of weary travelers, we follow Francisco's lead. Bill, the ER doctor, walks ahead with red knee-backs exposed to the sun. His son, David, totally recovered from his sore throat, leans against his father's shoulder. When we reach the end of our hike, Francisco plays volleyball with the crew on a high rocky surface. Pearl and Henry rest on a smooth stone ledge—sunlight streaking through a wave-sculpted archway at their backs. Overcome with sea and sun, the four of us collapse onto terraced lava steps—dangling our toes in icy waters until Francisco's "Time to go" punctuates the peace.

Down we climb, down to the dinghy awaiting us. Halfway there, a smiling Suzanne places her hands on my shoulders and pivots my body one hundred eighty degrees. "Mom, look at the bunnies," she whispers and points (I need to explain here that Suz refers to all beloved creatures, both animals and humans, as "bunnies"). There, above us on a cliff, a pair of blue-footed boobies dance and strut—their eyes focused only on each other. For more than twenty-five years I have dreamed this picture, and here it is. While our companions continue descending towards the dinghy, we stand transfixed, watching the pair until Francisco's booming voice forces us back to a world of clocks and schedules. As we motor away from the island, fish-craving boobies slice through the air, breaching the water's surface at tremendous speeds.

Later that evening, a few of us go out on deck to watch a spectacular display of constellations against the black night. Though with others, I feel strangely alone. With no city lights to distract me, no honking horns or blaring radios, I zoom through miles of earth's atmosphere toward a wide band of brilliantly blinking stars—our Milky Way. "Those two particularly bright stars," says a fellow gazer, "point to the Southern Cross. The brightest star at the base of the cross is *Acrux*." Following his directions, I find the star he names. I remember once asking my father what stars were made of. We were standing on a flat section of roof accessible through my bedroom window. It was summer and we were looking through his telescope. "Stars are made of gas, my little one, burning globes of exploding gas." "Like our stove?" I asked. "Much hotter, hotter than you can imagine." Straining, I search the equatorial sky for the Big Dipper as I'm used to seeing it: three stars on the handle, and four on the pitcher. Reversed in the Southern Hemisphere sky, the dipper looks like a four-sided kite with a streaming tail of lights.

Much seems reversed here, I'm discovering. I have withdrawn from my normal world—know nothing of the news nor of anyone beyond my family and companion travelers. Nothing is as it was, and yet, as Pearl says, I am home.

15
Walls Apart

Sunday, August 20
Morning in Genovesa, a Northern Island

With clouds of seabirds above, our *panga* skims gray-blue waters along the base of a sheer cliff, then pulls up flush against the rocks. We are in the northeast corner of the archipelago—farther north than most tourist ships travel.

No sandy shoreline to welcome us here. We grasp Francisco's hand, take hold of a wooden handrail, and immediately begin a steep climb up "Prince Philip's Steps," so called because His Majesty Prince Philip visited here in the 1960s.

As we wind upwards through a narrow gallery of stone, I run my fingers over an ever-changing exhibit of primal "art" created by wind, sea, and the "bunnies" that dwell here. My over-stimulated mind visualizes "paintings" in the guano—chalk-like designs in chasms and crannies and on time-weathered rock. I imagine the wispy spines of an abstract sea urchin; a primitive face with a frosting of hair over one eye; an etching of a tropicbird in flight, its long tail feathers like streamers of white silk. At the top of the stairs, a colony of red-footed boobies perches in the stark branches of *palo santo* trees. Unlike their blue-footed and orange-billed Nazca booby kin who nest on the ground, the red-foots nest

Red-footed Booby Bird

in trees and spend much time at sea—traveling up to four hundred miles to get food for a single chick.

Along the trail, the air smells of musky feathers. Galápagos doves with impressive blue eye-rings peck cactus seeds while Galápagos mockingbirds chase insects and feed on bits of carrion. "Most people think mainly of finches when they think of Darwin, but the mocking-thrush as he called it was an important piece of the evolutionary puzzle," Francisco tells us. "Darwin collected three of the four mockingbird species on the islands and because all three were so similar, he concluded they originated from a common ancestor."

Red-faced lava lizards, about five inches long, climb rocks and dash in and out of stone crevices. These lizards can live up to ten years, I remember reading. High above the cliffs, boobies with beaks down and wings pressed against their bodies, plunge into the sea with barely a splash—only to pop upwards seconds later with their prizes. Underwater photographers have proved they actually catch the fish on the way back up.

Great frigatebirds, with iridescent black wings and forked tails, scan the billowing waters for surface fish or squid. "Look at them," Francisco shouts. "Their bills are hooked, so they're not good divers, and their preen glands aren't big enough to do a good job of waterproofing their wings. But they are true survivors." We watch these amazing pirates await the exact moment when another's catch may become their own. Suddenly sky-borne, they pluck fish out of boobies' mouths with the tips of their bills, or catch falling rations before they hit water.

Continuing our walk, we reach a point where porous black rocks are overlain with black marine iguanas sending out sentries one after another into the water. Only the occasional sporadic movement of the sentries breaks the ebony expanse. "They're sometimes called 'imps of darkness,'" Francisco tells us, "but they're really quite harmless. I'm proud to say they are endemic to our islands and the only true marine lizards in the world." I remember Darwin's surprise when a seaman aboard his ship tied a line to one of them, sank it by attaching a weight to its body, and found it still very much alive when he pulled it up an hour later. Darwin found these creatures "hideous-looking" and "sluggish," but I find them strangely appealing—with their salt-frosted faces and powerful survival skills.

To the left of the iguanas, on a rock jutting out to sea, a short-eared owl uses his beak to rub oil into his feathers, then flutters his wings and ruffles his tail. Occasional clusters of yellow-tipped lava cacti, like hairy golden pickles, cling to bare stones where no other life appears. So rare and unexpected in this almost monochromatic land—the contrasting black and gold draw my attention and appreciation as they never would at home.

At home, Art is probably practicing his banjo, feeding the pets, or maybe talking across the fence to Ed and Tisa about what we'll do together when I return. If I'd heeded my internal warnings, I'd be there with him—happy, but missing all this. Lord knows, the "what if's" thundering in my head have kept me on the sidelines listening to friends tell about their trips often enough. And though I'm not one to sit around—for me there seems to be a direct ratio between worry time and procrastination. The more time I have for planning and worrying, the more likely I'll go dormant. On the other hand, if someone invites me on a last-minute mountain or desert hike or a middle-of-the-week movie, I drop what I'm doing and go. Or if business requires that I go to Sacramento or Los Angeles, I pack a few things and take off. Experiences like this one—the kind that require long-range planning—work me into a frenzy. I wonder how much joy I've missed throughout my dreaming years.

We go back to the ship to prepare for an interlude of snorkeling. Visions of hammerheads and white-tip sharks (neither of which are supposedly interested in people), manta rays, and stingrays overrun my thoughts. On this excursion, Francisco announces, we dinghy out to sea and dive "into the deep." My heart pounds, but I have little worry time and am determined to overcome my fear and see these creatures for myself. And so I head straight for our tiny bathroom to put on at least one disposable contact lens. Again and again, I try to place the gelatinous volcano-shaped blob onto my right eye. Again and again, it collapses and slips off the iris. I ask for Heather's help, which she earnestly tries to provide. But the mission remains unaccomplished.

"We have to hurry, Mom," Heather tells me. "Francisco is waiting for you."

Walls Apart

"Go on, Heathie. I don't want to keep them waiting." With this, I dismiss my chance of diving off the dinghy "into the deep." Disappointed *and* relieved, I urge Yolis to go with the other divers, but she insists the water is too cold and she has seen more colorful sea life in Cozumel. "Maybe, I could help you practice-dive right off the *Beluga*," she offers. Before I can object, she tells Alfonso and Reynaldo how much her cousin wants to snorkel, but how she just didn't get ready in time. "Would it be all right," she asks them, "if we were to dive off the yacht?" They pause, murmuring something about it being *muy hondo para ella aquí*, and then disappear.

Yolis tells me they are concerned the waters here are too deep for a beginner. Minutes later they return with the tall, smiling captain. *No es posible, Señoras*, he says gently. *Va a ser muy peligroso*—"too dangerous." A few minutes of Spanish pleading from Yolis, and the captain so regrets her poor cousin's loss of a once-in-a life-time dream that he directs Reynaldo: "Get the other dinghy and take them to the shore."

16
Darwin Bay without Guidance

Sunday, August 20
Later in the Morning on Genovesa

A willing Reynaldo motors us to the Darwin Bay landing on the southern side of Genovesa—a spot reserved by Francisco for our group's afternoon exploration. Knowing Francisco's meticulous adherence to schedules and rules, we feel concerned that he will discipline Reynaldo for our dalliance. When we tell Reynaldo of our concern, he grins: "It is the *captain's* decision, Señoras," he tells us in Spanish. "Whatever the *captain* says is law. Things will be fine, but you must remember not to walk on any paths. Just snorkel by the shore and I will come back for you in about an hour."

We are the only humans on this intimate, white coral beach in the Pacific. Sea lions, pelicans, and gulls sun themselves on the rocks and pay us no heed. Our solitary presence sparks a forbidden sense of adventure. We are here despite National Park rules that prohibit tourists without guides, but we are here with the captain's permission, and the captain's word is law.

This world we have entered seems all the more special because we are privileged guests. Gentle waves bathe our toes as we scan the cove and inhale the kelp-scented air. Charmed by all around us, we breathe more deeply, walk more slowly, and take in every nuance afforded us.

In a state of semi-rapture, we turn from the pristine shoreline to face forty- to fifty-foot cliff walls rimmed with native *Opuntia* cacti and mangroves. But unlike the walls of early morning that bore the imprints of nature—these walls are a testament from a darker side. "96 Remo" and "Moani" signal graffiti in paradise. Our Quito cab driver, Miguel, had talked about graffiti being communication for the powerless. Did the tourists who scrawled here, I wonder, want power over nature? Did they feel inadequate in the face of topography that had survived millions of years? Or were they arrogant enough to believe that they deserved their own monument? Reminded of the dangers these islands face from our own kind, we retrace our steps through the damp sand until we reach the sea.

I am here to snorkel, and snorkel I will. To give me the "full experience," Yolis lends me her fins since she has no plans to insert more than her toes into the cold waters. Spreading her towel on the warm sand, she plunks down her backpack and camera. I sit at the water's edge pulling on my fins, then inch my body into a standing position. Like a clown with oversized feet, I rock sideways, tilting so far in one direction that I crumple onto the sand. Yolis runs to prop me up, and once again I stand—bobbing back and forth. I lose my footing over and over—with Yolis clinging to my wet suit to help me stay upright. Losing control of our bodies and minds, we collapse together into the cold waters unable to stop an almost primitive burst of hysterical laughter.

I finally decide against a "full experience," replace the fins with shower shoes, and snorkel close to the shore. Through my blurred, uncorrected eyes, I see a blend of sandy water, gray fish, and lots of urchins and anemones. From the shore, Yolis takes documentary photos of my goggle head with yellow snorkel tube extended. "Now we can prove you really did it," she calls.

At around noon, I trudge out of the water and sit with my cousin on the white coral sand. Back in the states at Seal Beach, California, my niece Kaitlyn is getting ready for a total-immersion baptism. "Happy baptism, Kaitlyn," my cousin and I holler—with sea lions our only witnesses.

I think of evolution and creationism and how the conflict between them never ends. Darwin himself feared recrimination and once dreamed of being hanged for heresy just for going public with his research. There are probably a few guests at the baptism who might think I'm an atheist just because I'm here. To me, evolution has nothing to do with whether or not God exists or whether God created the world. It deals with scientific evidence. And even though scientists have proved that the universe has been changing for billions of years, many still believe that the earth and all its inhabitants, including humans, look exactly as God created them within the past ten thousand years. I really don't understand the conflict. As I see it, an all-powerful God could have created the world any way He chose.

"Look, Bette, he's coming," Yolis calls out. Just a glimpse of Reynaldo aboard his white *panga* motoring toward us, and my thoughts segue from science to fun. He is grateful that we have not trespassed the trails and pleased that I got some practice time.

Lunchtime on the Beluga is filled with conversation and delicious *llapingachos,* inch-thick potato pancakes with scallions and cheese. Those who dove off the dinghy while we were on our private adventure rave about the sharks and the turtles. Those who stayed on board the ship rave about their naps in the peace and quiet. Neither Yolis nor I say anything of our interlude, not wanting to compromise Reynaldo or the captain. I wonder if Francisco knows we've strayed.

17
Darwin Bay with Guidance

Sunday, August 20
Afternoon in Bahía Darwin

Later that afternoon, as we motor to Darwin Bay with our fellow voyagers, I think about this morning. I watch Francisco chatting with Deborah and Matthew. As guardian of these islands, he would be justifiably angry at our having been here without supervision. Yes, it was fun—something to talk about with friends, but I don't think I'd do it again. Park regulations, after all, insure the safety of the Galápagos and the beings I love.

Back on the white coral beach of Bahía Darwin, Yolis and I wait for Francisco to speak about the graffiti. But he disregards it until I ask. "The earliest graffiti," he tells us, "was left in the seventeenth or eighteenth centuries by buccaneers and pirates. It is now covered with overgrowth. All the graffiti we see was done in guano, not paint," he emphasizes. He says nothing of "96 Remo" or the other "new" entries. And though I want to know why these messages remain when they could be easily flushed from the rocks—something in his silence convinces me to ask no further questions.

Past tide pools bordered with lava gulls and swallow-tailed gulls, we take the trail forbidden us during our morning visit. Entering a mangrove forest, we see great frigatebirds perched

Nazca Booby Overlooking Darwin Bay

in saltbush, their iridescent feathers gleaming in the sunlight. A colony of red-footed boobies roosts on high and low hanging mangrove branches. The green trees contrast sharply with the black lava and the boobies themselves—with their vibrant feet and yellow eyes lined in Maybelline® blue.

The trail leads us over jagged lava stones that jiggle precariously. We soon discover that a smooth walking rhythm is impossible, for the intervals between stones are irregular. Some stones are flush and connected, others inches apart, and still others require muscle-stretching steps up or across. Pearl confesses that one of her big fears is walking on unsteady rocks, not knowing if one will shimmy and cause her to loose her balance. Her fear of falling keeps her at the end of our line so that she can follow the path made by others. Francisco, up front with the more daring, walks rapidly. I offer Pearl my help but she's sure that "having someone close by is enough." She follows me slowly as I try out the steadiness of the stones. "It'll be O.K," I assure her. "You're

Darwin Bay with Guidance

Talking to a Galápagos Sea Lion

doing fine." With her eyes cast downward on the rocks, she mirrors my steps.

During our trek across the lava, Yolis takes my place guiding Pearl. I walk ahead and lose track of them until Yolis calls out that Pearl has fallen. Heading back, I comfort my friend who is badly shaken but all right. Her backwards fall could have been severe, especially with her slight build. I can only imagine what might have happened if her soft backpack hadn't cushioned her fall. Supported by friends, Pearl perseveres along the trail, talking of other trips she's taken—like the one to the Amazon in which guides stayed with or assigned others to help seniors and others who were unsure or needed assistance. Francisco is far ahead and knows nothing of the accident.

When we return to the bay, we watch two swallow-tailed gulls deliberating on the perfect piece of coral to add to their nest. The female picks up several "identical" looking shards and discards them with her mate's blessing. Using a selection process that humans can only guess about, the two finally agree on a piece and add the chosen adornment.

The beach is much hotter than it was this morning. Thanks to layered clothing, I peel down to my bathing suit and enter the brisk waters toes first. Inching forward, bending until my knees, waist, and shoulders are covered, I roll over and over until totally immersed. Pearl and Yolis join me, and we swim together just feet from bellowing sea lions. Behind us, Heather and Suzanne explore the cliffs, and five-year-old Cathy prances along the shore. To our right, Dr. Bill cavorts with a playful sea lion calf while Jennifer and David cheer him on. Perched on a rock five feet away, a lone brown pelican withdraws into his world of feathers and fins, a world we may transverse but never call our own.

18
Moon Landing

Monday, August 21
Morning on Bartolomé, a Central Island

"Remember to stay on the trail," Francisco calls out as our *panga* motors close to a carved stairway rising just above the water. We disembark smoothly and begin a steep but easy climb overlooking tide pools. Tow-headed Cathy tugs on her mother's sweater: "Look there," she squeals. "More blue-foots—they're so silly." My eyes follow her tiny finger up, up where the blue-footed ones strut along the cliff edge. "Spanish sailors called them *bobos* or clowns," her mother tells her, "because they make people laugh."

"Walk single file," Francisco directs. "You are now on an extinct volcano. This gritty, blue-black lava ash comes from molten basaltic rock that once flowed across this island. The ground is soft and fragile and will disintegrate and blow away if you walk on it." Determined to be respectful human visitors, we stay on the narrow path, and in a straight line cross the charred, cratered landscape under a metallic gray, rain-threatening sky. Through the mist I see whimsical silver-gray *tiquilia* plants that elsewhere grow thick and mat-like, but here, like flimsy bushes, dig their roots deep into the ash. Unlike the other islands we've visited, Bartolomé brandishes dramatic reminders of how these islands came to be.

Volcanic devastation is everywhere: spatter cones or steep hills formed by the splattering and hardening of boiling lava; broken lava tubes formed by rivers of fast-flowing melted rock; tuff cones of hardened volcanic ash; volcanic bombs—fragments of light ash, some measuring three to four feet across, lying like immense boulders that appear immovable. At Francisco's urging, we take turns raising the three-to-ten-pound "boulders" above our shoulders while members of our families capture our "incredible strength" on film.

As we progress, the trail becomes a cedrela wood stairway built by the Galápagos National Park to protect the land from erosion caused by visitors. We climb three hundred eighty-four steep stairs, resting for heart-pounding moments on platforms for the weary. Beyond the first set of stairs, I notice yellow-tipped lava cacti shaped like tubes and growing in clusters. These are pioneering plants, we learn, because they are often the first to establish themselves in this soil-less earth. Like lichen nourished by the minerals in the lava, this endemic species imperceptibly converts rock into soil.

After a half-hour climb, we reach the summit and the Foralete Bartolomé Monument. Francisco faces us, standing strong against the wind—his dark, neck-length hair flattened against his head. "This is one of the most often photographed views in the Galápagos," he tells us. "And it's one that Darwin never saw—since he never stood atop this small island. During his five-week stay, he visited only four islands: Santiago, Isabela, Floreana, and San Cristóbal."

Standing on the precipice with my family, I look out onto Pinnacle Rock, a vertical tower of hardened ash jutting out of the sapphire-colored Sullivan Bay. Moored in the glassy waters is a scattering of islands. One by one Francisco points and calls each island by name: "Santa Cruz, Santiago, Baltra,

Moon Landing

View of Pinacle Rock from Bartolomé

North Seymour, Daphne Minor, and Daphne Major where for twenty years Princeton scientists Peter and Rosemary Grant have been studying the continuing evolution of Darwin's finches." Heather touches my shoulder: "Mom, that's the study I was reading about on the plane."

Seemingly peaceful and still, these islands ferment with what is to come. They have erupted more than fifty times in the last two hundred years, Francisco tells us.[1] "In 1954, on the coast of Isabela Island, a coral reef suddenly rose fifteen feet from below the water—extending the shoreline about three-quarters of a mile and exposing sea turtles, eels, crabs, and fish of all kinds to the murderous rays of the sun. There's a story told around here, maybe true, maybe not, that some fishermen dropped anchor in Urvina Bay before the uplift.

---

[1] The most recent eruptions have occurred on Isabela Island at Cerro Azul Volcano (May 2008) and Sierra Negra Volcano (October 2005), and on Fernandina's La Cumbre Volcano (May 2005).

When they returned for their boat, they found it on land." I picture the poor fishermen expecting to find their boat in the harbor where they'd left it and instead finding it grounded and surrounded by dying and dead star fish and lobsters. When Francisco's voice pierces my reverie, my mind circles back to him.

"In 1995, when I was assisting on a photo shoot in Fernandina, I saw a dramatic sky which turned out to be from an eruption on Cape Hammond only miles away. I didn't hear anything, just saw the red sky," he recalls, "but the power of the explosion discharged a river of lava five kilometers or about three miles—all the way to the sea."

As I listen, I can almost sense the earth's shifting forces. Someday these islands will disappear into deep ocean waters and others will appear, supporting a continuous evolution of creatures and formations I will never know. I stare at the steep and jagged Pinnacle Rock[2] standing sentry over the lands and the slick waters that separate them. Above this needle of a rock, frigatebirds, their expansive wings quivering, hover in a delicate balance above their now-burgeoning food chain.

Down three hundred eighty-four stairs, we hike back to the landing with its tide pools and black marine iguanas that seem one with the cliff walls they climb. Yellow plastic bottle caps and bits of tissue float and cling to lava rocks. I think of human wastefulness and how nature wastes so very little. What one organism casts off or leaves behind, another uses—whether

---

2 During World War II, the island of Baltra, also called South Seymour, served as a U.S Air Force Base to protect the Panama Canal. U.S. pilots used Pinnacle Rock for target practice, probably because no humans lived in the vicinity and because it was easy to see. Remains of an artillery shell can still be found in the upper part of the rock.

Portrait of a Galápagos Penguin

seeds, plant debris, or the remains of a seal pup. I strain to retrieve the scraps, but they are beyond my reach.

We hustle into our *panga*, take Francisco's extended hand, and slide into place—four on one side and four on the other. The spinning whirr of the motor sounds, and we skim across choppy gray waters in search of Galápagos penguins who play sporadically on the rocky shore of Pinnacle Rock. "Watch inside the caves," he tells us as we near land. "They nest there. Sometimes, we're lucky and sometimes...."

"Mom, look," Suzanne calls. Slick black feathers against the rocks are, at first, a camouflage too perfect to penetrate, but as my eyes sharpen, the flightless birds zoom into focus. I see three of them—two black and one gray—no more than fourteen inches high. One penguin turns towards the *panga* exposing stark white belly feathers. Wings outstretched, all three face into the wind to cool themselves, a maneuver that has served them well—probably since the ice age when their ancestors arrived on the waves of a Patagonian storm. No one knows their exact origin, but the cold Humboldt Current from the South Polar Region transforms these equatorial waters into a perfect home. Two penguins stand together preening feather layers that are thinner than their Arctic brothers'. A black one waddles to a ledge covered with red sally lightfoot crabs. Nearby, a brown pelican scoops dinner from the waves while a diving booby hits the water. The penguins pay no attention.

The ravages of the last El Niño and the ever-growing human presence on these islands have burdened these fragile creatures with much to overcome. We have seen what we've come to see and our *panga*, bobbing in the swells, must go—leaving the penguins to their enduring struggle.

19
Discord in Paradise

Monday, August 21
Early Afternoon in Sullivan Bay along the
Coasts of Santiago and Bartolomé

Back on the *Beluga*, we get ready for snorkeling. Yolis and I select the calmer northern beach off Bartolomé while Suzanne, Heather, and the more adventurous swimmers will again dive off Francisco's dinghy in search of hammerheads, white-tip reef sharks, and rays. With Francisco's permission, Reynaldo motors the two of us to a beach overlooking Sullivan Bay. But unlike our solitary adventure on Genovesa, this adventure we will have to share with travelers from two other boats. Respectful of the regulations given by Reynaldo, we know we must never stand on or touch the lava rock framing the left side of the beach. "And don't forget," he cautions, "you can snorkel and sit on the sand, but don't go near the sea lion mangrove shelter area." Before leaving us on our own, he asks a guide from another group to "watch over us" and then dinghies away.

The cove seems the perfect place to snorkel, with its intricate underwater rocks and ridges, and its proximity to shore and Pinnacle Rock. Immersed in the chilly water, I'm grateful for the wet suit even though it's sleeveless and has a scoop neckline. I lull myself with the echoing sound of my own breathing in and out the tube as I pass over white-banded angelfish with fan-shaped orange tails. A school of yellow-tailed surgeonfish flashes by—razor-sharp spines jutting out from the base of their tails. These "surgical" spines, I've read, protect them from attacking predators. My view of sea life is as unobscured as through the windowpanes of aquariums back home. But here, the current's constant pressure reminds me that I am not separated by glass, but am one of many creatures skimming above the ocean floor.

Two blue parrotfish and a large school of creole guide me through the liquid maze until an unseen presence disturbs the peace, and I surface. A few feet away, a tourist slashes and kicks his way through the water oblivious of me and the sand he displaces. Startled by the splashing, a sea lion darts toward a safer haven while a pelican, unfazed by the disturbance, gazes down at me from a rock. I confide in him my awe of his world, then continue facedown in the water. Minutes later, despite my strong attempts to avoid the jagged rocks, the current pushes against me and I press the heel of my right hand onto a lava spike. I raise my head and see that I am headed for a labyrinth of caves around Pinnacle Rock. The beach behind me, though relatively close, seems distant, and I can barely discern the outline of my cousin lying on the sand. Having seen enough, I swim back to shore—grateful for not having braved the powerful current on the other side of the beach where my girls now float with manta rays and hammerheads.

As I walk up the warm sand toward Yolis, I'm stunned by the appearance of the mangroves behind her. Curtains of yellow, blue, and red towels drape across low-hanging limbs. Cabaña hats, printed shirts, and snorkeling gear transform this once-pristine environs into a market place. Five women

Discord in Paradise

Profile of a Pelican

sit laughing below the evergreen branches—inches from a family of sea lions. Their guide who promised to "watch over us" has disappeared. We call to the tourists in Spanish and English: "Please—you must not hang clothes on the mangroves. You must respect the space of the sea lions." They pay no attention. Below us, a woman stands on the fragile lava rocks we have been told to avoid. She raises one foot, shakes off sand, then raises the other. A man holds her shoes and hands each to her when she bids. Yolis calls out in Spanish: *Señor, no puedes parar allí. Es contra la ley.* "You can't stand there. It's against the law." He repeats our warning to the woman who laughs and walks from lava rock to rock. How can this happen, we wonder. Where are the guides? What of the national park's vow to protect these precious isles?

Later, when the *panga* returns, we tell an attentive Francisco what we've seen. He scans the mangroves where the guide has rejoined his flock, then leans toward us—his eyes narrow and his forehead furrowed. "Not all take our mission to protect so seriously. I cannot talk to the guide in front of others, but I will see him in private." We know he will.

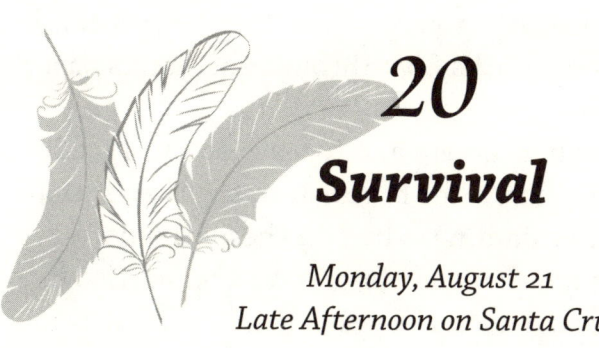

20
Survival

Monday, August 21
Late Afternoon on Santa Cruz

In just four days, I have grown accustomed to change on these islands. I have walked among giant tortoises once numerous—now protected, listened to Francisco's story about the sudden uplift on Isabela, and witnessed volcanic devastation on Bartólome. Not only have I learned to accept change. I now expect it.

The weather is cool and overcast, in striking contrast with the warmth of Sullivan Bay just hours ago. Navigating the waters of Conway Bay, our boat approaches *Cerro Dragón* or Dragon Hill on the northwestern side of Santa Cruz—a totally different landscape from the southern parts of this island we walked our first day.

Here we will see no *Scalesia* forests, pit-craters, or non-native papayas and bananas. We will not walk cobblestone streets or stop to finger the wares of villagers eager to please, eager to sell. Instead, we hike along a path bordered by leafless gray *palo santo* or "holy stick" trees. When cut, Francisco tells us, their branches exude sweet-smelling incense often used for religious ceremonies in Ecuadorian churches. I lower my head slightly and take a sniff. "Mom, they have to be cut," Suzanne teases. "Just checking," I tell her. Minutes later, we reach a lagoon where rare lava gulls, known for their pilfering talents, scavenge for crustaceans. A few feet away, juvenile white-cheeked pintail ducks float through the waters past a wading common stilt.

Heather steadies her camera and focuses on one lone flamingo dipping its bill into the brackish water. "Fewer than five hundred greater flamingos live on these islands," says Francisco. "Notice how pink he is, not coral and mottled like the ones you see in your zoos. Here, they get the reddish brine shrimp they need for their true color. And the brine shrimp our flamingos eat are red because they eat reddish algae nurtured by these waters. " I sense both pride and smugness in his words—that here on his islands these lovely birds get what nature intended.

Taking leave of the water dwellers, we wind our way through the lowlands where long ago, before sailors and whalers decimated their community, tortoises roamed freely. Today, a tiny population lives in the area, and others from the highlands migrate here in February and early March. Arid and warm, this landscape is dominated by the giant prickly pear cactus or *Opuntia*, a paradigm of adaptability. To retain moisture, these plants actually close their pores in the heat of day.

"On islands with tortoises," Francisco tells us, "young cacti, which are especially appealing to tortoises, have evolved ultra-thick spines to protect themselves." When the cactus grows, he continues, and its pads are beyond a tortoise's reach, its spines soften and the bark itself deters the gentle giants. How very ingenious, I think, and then remember a Darwin quote I read recently: "It's not the strongest of the species that survive," he said, "nor the most intelligent, but the one most receptive to change."

Survival

Male Magnificent Frigate in Search of Mate

The rocky, uneven trail leads through a thick green mat of *sesuvium* with small, reddish, star-shaped flowers. Along the way we pass a plant with fuzzy, sharp-pointed leaves and fruit resembling small tomatoes. "This is our Galápagos tomato," Francisco emphasizes. "Our islands were its only home until U.S. scientists took their DNA and crossbred them into a salt- and drought-resistant cherry tomato. Now, it's in danger of losing its endemic status."

I am puzzled, but don't question him. From what I know, an endemic species can't lose its status unless it colonizes somewhere else naturally. I glance at Yolis to see if she too feels his recrimination. She returns my glance and smiles knowingly. My cousin and I sense Francisco's devotion to his islands and his anger that foreigners have taken advantage of what belongs to his people. As a writer I can imagine how I would feel if someone misquoted my words and used them for purposes I had never intended—even if those purposes were worthy. Drought-resistant cherry tomatoes may be desirable to others, but to Francisco, they are symbols of how his islands have been violated. Something precious to him has been recast by strangers in sterile laboratories far away.

Onward we walk through an impressionistic forest of silvery *palo santo*. Larger than any we have seen on other islands, these sacred trees await December rains which call forth leaves and cream-colored blooms. Unfortunately, we will not be here to see them. The trail leads past mocking birds in search of locusts and lizards, past yellow warblers aglow in silver branches. Brown flycatchers with mustard-colored bellies pluck insects in flight while a cattle egret perches motionless in a nearby tree awaiting the perfect invertebrate meal.

The pathway becomes repetitive and we chatter among ourselves until Francisco's "Quiet" hisses through our line. Instantly we comply and wait for his directions. "Stay on the path. We are approaching land iguanas. They are skittish. Back in 1976, wild dogs running in packs almost wiped out their entire endemic population from *Cerro Dragón* to Conway Bay."

I freeze in place inches from a mature cactus. At its base I see a reddish-orange pair of iguanas munching pads and thorns. They are about three feet long and have tracking numbers on their backs. Hypnotized by their movements and intermittent stillness, I admire their leathery dorsal crests and body folds, their pointed noses—unlike their flat-nosed marine kin—and their perpetual smiles.

When we are several yards beyond the pair, Francisco tells us about a fantastic iguana rescue operation. After the dogs decimated the iguana population, teams from the Charles Darwin Research Station and the Galápagos National Park Service rescued sixty of the poor creatures. The research

Survival

Galápagos Land Iguana

station had enough cages for twenty-two iguanas, and so thirty-eight were placed in "semi-captivity" on *Venecia*, a small island just north of where we are now. Wardens from the park service transported ninety tons of earth from *Cerro Dragon* to the mostly lava island and built an artificial nesting area. For more than ten years, devout conservationists nurtured both populations until in 1987 the iguanas were ready to go home. And so their recovery continues to this day.

Silently I follow wavy, tail-drawn etchings in the earth where these amazing reptiles have walked. A lone iguana inside a shallow tunnel dug for shelter and nesting is crunching a prickly cactus pad, devouring spines and all. I recall my uncle tweezing thorns out of my eight-year-old tongue after I'd taste-tested a cactus apple. My hungry iguana friend needs no such help. If a thorn sticks in her tongue, she can remove it by simply rubbing her tongue back and forth against her teeth. I watch as she pauses—savoring this *Opuntia* delicacy that provides both food and water.

Leaving the iguanas, I walk quietly with my daughters and cousin and think about change. When Darwin visited the Galápagos, he found iguanas so numerous on Santiago Island he couldn't find a spot free from their burrows. Today, land iguanas are extinct on Santiago and endangered on other islands. I remember the twenty-six year old Darwin's repulsion at their "singularly stupid appearance." While he found them "ugly," I find them enchanting. But I am mature, and he was very young.

21
"Valentine" Island

Tuesday, August 22, 2000
Morning on North Seymour, a Central Island

It is the fate of most voyagers, no sooner to discover what is most interesting in any locality, than they are hurried from it.—Charles Darwin

It is the fifth day—our last morning on these islands. In just a few hours, we will be back on the Ecuadorian mainland. I am keenly aware of the spray from the sea, now as familiar as air. My senses have grown so accustomed to the smells and sounds of this sanctuary that thoughts of separation loom heavily, and I must strain to push the gloom away. After waiting for swells to subside, we take short leaps from our boat onto slippery black boulders dotted with guano from gulls, pelicans, boobies, and frigatebirds.

A short climb to a comparatively flat, sandy area and we are greeted with wild activity that more than offsets the colorless monotony of the landscape. Sea lions and their pups cavort in a rookery alive with nursing, barking, and nuzzling. Speckled with white sand, pups chase after their mothers and siblings, and block our path. Pausing for seconds to listen and respond in our limited but appreciative human way, we take care to walk around them. Undaunted by our presence, several snooze on the warm sand, their flippers gently palpitating. One pup, not more than a few weeks old, rolls over and watches me with soft, dark eyes.

"Mom!" Suzanne almost shrieks. "Turn around." I pivot in her direction and almost lose my breath. There just a few feet away, a scene captured in hundreds of nature films materializes before me. Eyes and beaks to the sky, wings held back against soft bodies—honking female blue-footed boobies and their whistling suitors raise their lovely blues at forty-five-degree angles, first one then the other. "Look at my feet, look at my beautiful feet," they seem to say, oblivious of sea lions frolicking among them. Immobile, I watch their wing-flapping dance of love, their mysterious exchange of *palo santo* twigs as symbols of commitment—twigs that will never be used for their nests. "Listen to the voices of the males," says Francisco. "They sound a bit like asthmatic sopranos, don't they? Now, notice how their eyes look smaller than their mates', but don't be fooled. See that dark pigment ring around the eyes of the females. It's like a makeup trick to make their eyes seem larger."

Hiking along, we avoid stepping into circles outlined in chalky guano on the bare ground. These "nests" atop the volcanic soil seem basic and austere, without a single twig or leaf to adorn them. Several ringed enclosures have feathery occupants warming eggs between their azure feet. Though I tiptoe past not to disturb them, they seem unaware of foreigners trespassing their territory. The thought of blue-foots creating circular nurseries in the middle of the trail makes me smile. Do they orbit—pooping as they go until the rings are complete, I wonder? Or do they poop their nest boundaries, one plop at a time, over days?

"Valentine" Island

Mother and Baby Blue-footed Booby

Francisco explains that unlike their red-footed relatives, the blue-foots can successfully care for three chicks instead of one. When the babes are hatched, both males and females attend them, and for the first four months, either the mother or father is always present. In one circle, we see a fluffy white chick stumble while a doting parent watches. In another, a feeding parent opens wide while a sweet babe leans far into the beak for food. As we walk along, we must also take in and accept somber scenes of dying and deserted young, too frail for survival, their cottony white feathers fluttering in the breeze. These scenes, I now know, are as much a part of the ever-changing Galápagos as those of thriving life and newness.

Beyond the boobies, we watch the magnificent frigates with their pterodactyl bodies and iridescent purple shoulder feathers. Wingspans that can reach ninety inches make these birds superior air-racers. Their pirating skills, now sidelined in the hoopla of passion, have earned these amazing aerialists reputations as "man o' war" birds.

In the branches of the *palo santo* and in the skies, I see them—black, bat-winged creatures with scissor-like tails. Females fly above, surveying and selecting from among tree-bound males displaying balloon-like vermilion throat pouches or "gular sacs." What draws a female to one particular male, I wonder. Does size matter? Subtle variations in color? Shape? From my vantage point, all the gular sacs look alike. At times two males fight, and one punctures the other's pouch. Luckily, the pouches heal.

I listen as hopeful males grouped together on branches make rhythmic drumming sounds while wagging their heads and vibrating their wings. They lean back on their tails, arch their heads, and thrust out their chests to give females the best possible view of their pouches. When their magic works, females drop from the skies and the mating ritual begins. We can't stay to watch the bonding behavior, but through Francisco's description I visualize shaking heads, clattering beaks, and the coiling of delicate necks around each other.

Unlike the blue-foots, he tells us, frigates build scraggly nests on tops of bushes and small trees. The male fetches construction materials while his mate builds the nest and lays a single egg that must survive up to a 50-day incubation. Both the male and female watch over the nest, taking turns flying out to sea for food until the chick's about three months old. Then the male disappears, leaving his mate to care for the juvenile. Parenthood for a frigate female is demanding. Her babe begins to fly at about five months of age, but she continues nurturing it for nine or more months. Rearing a chick takes so long that female frigatebirds can't breed every year.

In spite of vigilant care by the parents during the nesting and hatching period, the survival rate of offspring is low, with eggs and hatchlings often slipping through the fragile nests to the ground. Sometimes, other frigatebirds snatch twigs, causing eggs or youngsters to fall. When offspring die, the parents choose other mates and begin the process all over again.

I feel relieved to see no signs of death or dying here. What we do see are parents sitting patiently with fluffy white chicks who look nothing like them except for hook-tipped beaks. It's hard to believe that these quiet creatures nurturing young could ever swoop through the skies pilfering from the mouths of hungry neighbors. And yet I've witnessed these survival skills myself.

We continue along the path toward the sea—past leafless *palo verde* trees with curved stalks and spiked spines that spell danger to fledgling boobies and frigates—and along cliff sides alive with the jade green succulence of seaside heliotrope. Because we're at the end of our stay on this and all the islands, I ask Francisco if he will take pictures of our group. Shoulders draped with camera straps from nearly everyone aboard, he clicks thirteen shutters, waiting for smiles and hugs and repositioning. More emotional than usual, he talks of a lucky incident involving North Seymour that kept land iguanas from going extinct on Baltra, also known as South Seymour. "We haven't seen any land iguanas here on North Seymour today," he tells us, "but they do exist on this island because of something that happened long ago."

One foot atop a boulder, he gestures southward toward Baltra, the island of my imminent departure, and recaps the story of the Hancock expedition that took place in the thirties. A group of scientists and adventurers collecting specimens for a California zoo and museums discovered that land iguanas on Baltra were plentiful but thin, probably because of goats destroying their vegetation. On North Seymour, where the vegetation was plentiful, they found no iguanas at all. Curious as to how iguanas would fare in a new location, they transported about seventy from Baltra to this island.

In the meantime, WWII broke out and Ecuador gave the U.S. permission to build an air base on Baltra because of its proximity to the Panama Canal. Dynamite blasted the lava to prepare for airstrip and building construction, and the sounds of hammers, drills, and voices replaced stillness. Soon Baltra's iguanas and other creatures were sharing their home with about a thousand military personnel. But people weren't the only foreigners on Baltra. Wild goats, dogs and rats continued to prowl the land, devastating the habitat. By the end of the war, the iguana population on Baltra had died out, probably due to the presence of introduced animals and people.

In contrast, the adult iguana population on North Seymour persisted, but the survival of their young was in jeopardy. To remedy the situation, conservationists transported some to the Charles Darwin Research Station in 1980 and by 1991, as a result of a rigorous breeding program, began returning them to their home.

Today, iguanas again crawl the Baltra landscape. Lucky visitors may see an occasional one close to the airport. Iguanas are also doing well here on North Seymour—which many of us, inspired by the love-making sea birds, are calling "Valentine Island." They live on the other side of the island,

"Valentine" Island

Francisco tells us—away from tourists. Smart creatures, I think.

Two dinghies bob in the waves. One will take its passengers back to the *Beluga* for lunch, to Floreana for the afternoon, and to Española and Turtle Cove in the days to follow. The less fortunate among us must take the other dinghy back to Baltra where we will board a plane for Quito.

Once more, I reach out to Francisco, slide into place, and wave to those I will never see again. I think especially of Pearl. We have exchanged e-mail addresses, but because we live on opposite coasts, a friendship will be difficult to sustain. Breathing in the musty, salt-perfumed air—I wrap my own arms about me to insulate memories of the last five days. Unlike Darwin I have made no brilliant discoveries, will write no scientific report that will change the world. Yet I myself am changed. With my family I will head back to the excess clothing, shoes, and medical supplies stuffed into plastic bags and left with Marcía at the Orange Guest House in Quito, back still farther to California and the life I have built there. But the way I view that life will transform because of what I have seen here.

This journey has affirmed my connectedness with other beings on this earth. I have stood inches away from creatures that exist nowhere else—creatures less afraid of humans than we are of each other—their trust upheld by national park regulations. In other parts of the world, there are no laws to protect innocent life, or laws so weak or enforcement loopholes so egregious that the breakdown in trust among living things is inevitable. What ultimately happens to this shared universe becomes the responsibility of those who care, and the legacy, for good or evil, of all who will ever live.

I am not the "most intelligent" or the "strongest," but as one "most receptive to change" must summon others to change for survival sake. In the past I could watch a newscast about an oil spill, or a river dying from toxic waste and feel genuinely moved—then turn off the TV and go back to whatever I was doing. I won't be able to do that now because I have evolved and because these atrocities are no longer abstract. They are personal.

I recall Francisco's words as he cautioned us to remain a respectful distance from the sea lions on Rábida: "They may touch us, but we must never touch them."

Epilogue: Full Circle

We didn't go home right away. We reunited with Nelson and Miguel for more sightseeing in Quito and accompanied Suzanne to the airport for her journey back to her teaching job in Oakland. Heather, Yolis, and I flew to Cuzco, Peru, and then on to the Sacred City of the Incas hidden in a crook between two mountains in the Andes: *Machu Picchu* "Old Peak" and *Huayna Picchu* "Young Peak." From the ancient city towering above the Urubamba River cloud forest, we climbed about three thousand feet to *Huayna Picchu*, looking down on terraced temples, palaces, and dwellings mysteriously abandoned almost five hundred years ago. In the days to follow, we zigzagged by rail and dusty taxi through the Sacred Valley into the village of Pisac with its precision-built Inca masonry and marketplace filled with fine pottery and woolens.

Returning to my California world with its deadlines, computer viruses, and traffic congestion was a shock at first, for I wouldn't or couldn't let go of all that I'd seen. Weeks after my return, my husband and I went to a local restaurant adorned with a seventy-gallon aquarium. While waiting for our meal, I watched bobbing seahorses and a blue parrotfish biting a chunk of coral with its beak-like mouth. Only weeks ago, I had swum with members of his family off Rábida and in Sullivan Bay. A graceful anemone waved long white tentacles to its orange and white clownfish companion, and a black angel darted among sunny yellow butterfly fish, a powder blue tang, and several other "reef" dwellers. All were cramped into a decorative, fluorescent environment for our pleasure. I have enjoyed viewing aquariums for years, but this time I felt sad. These creatures, bred and raised within Plexiglas®, have never known the world for which they were created—a world I have been privileged to enter.

A boy about eight years old approached the tank and began knuckling the glass to attract the attention of an orange-brown and white striped fish with an elegant "mane" and feathery fins. The fish was slow moving and grand; the child exuberant and noisy. Without hesitation, I left my seat and approached him: "You must stop that," I said firmly, but gently. "He is a lionfish—a king just like the Lion King. I know you would never want to disturb the Lion King." The boy stopped and eyed me with interest, not disrespect. "Mom," he called out to a woman already making her way to "rescue" him from my meddlesome tongue. "Mom—he's a lionfish. He's a king." I smiled at the two and went back to my husband.

A few months later, I took my elderly mother-in-law to the San Diego Flood Control Channel to watch sea birds. The wind, like the wind at Foralete Monument, buffeted us, but we didn't mind because we were focused on snowy egrets agitating the shallow bay waters with their bright yellow-sandaled feet. "They're a lot busier than the cattle egrets we saw on the Galápagos," I told her. "Oh," she said, "what a wondrous adventure you had."

Since my return, I've noticed that I don't enjoy zoos the way I used to because I now imagine the "inmates" in their natural state. I appreciate and respect the education and conservation roles zoos play, but prefer learning about wildlife by

Epilogue: Full Circle

watching the Discovery Channel or PBS specials. Still, when my visiting nephew Paul and his family asked if I would meet them at our famous San Diego Zoo, I told them I would wear pink and stand on one leg in front of the flamingo enclosure. Arriving in bright pink, I clashed with the coral-orange birds on display. I visualized Francisco flashing an "I told you so" smile and reminding me that in his world flamingos are pink because they get the diet nature intended.

Later that day, we watched an Andean condor—wings outstretched ten feet from tip to tip—fly the width of his cage, then break flight abruptly—landing on a tree limb to avoid crashing against his metallic restraint. I cringed and moved quickly past his cage—picturing his sovereign Peruvian kin rising on warm air currents thousands of feet above the Urubamba River Valley.

As time passes, I find myself increasingly intrigued by animals in their native environments. Things I've never noticed before, I now study and celebrate. The pleasure of my trip reverberates each time I look through the binoculars my husband bought me. For months, Art and I checked the progress of a red-shouldered hawk pair nesting in a thirty-foot cottonwood tree next door. Not a day passes when we don't see wild mallards dropping from the sky into my neighbor's pond; or crows in flight, pecans in their bills, landing on phone wires to feast.

When Pacific-coast flycatchers built a scraggly nest inside a light fixture on our front porch, we stopped using the light and delayed our house painting project. Unlike the blue-foots and the frigatebirds, the mother bird tended the eggs by herself. When they hatched, though, like her Galápagos peers, she shared the parenting role with her mate. About six inches long, with olive-yellow bellies and white-ringed brown eyes, the parents were either on the nest, catching food in mid-air, or doing sentry duty atop a stake in our geranium pot.

One Thursday in July, we discovered the flycatcher family had gone. That Saturday, Art dismantled the nest to get ready for the painters. By the following Tuesday, the pair had built another nest—this time on a light fixture next to our garage door. We put our painting plan on hold, and again stopped using the light. "They're double clutching," our friend Wayne told us. About a month later, despite our close watch, they disappeared without our seeing a single flying lesson or their winged departure beyond the pines and over the brick wall.

The flycatchers and all the sweet creatures on our road remind me of where I've been and all that I've seen. I think a lot about the fears that kept me from making this trip: the small, rickety planes and insurmountable physical obstacles that never materialized; the violence that did but was non-life-threatening. For twenty years, I convinced myself I had neither the courage nor the spirit to take such a trip. I consigned my dream to my daughters and watched it from afar—gathering its silent momentum.

Recently, when Art's mom announced that one of her biggest regrets was "never having had a love affair," I wondered what unfulfilled fantasies I would look back upon in my nineties. I know I won't be able to create fairytale endings for all my dreams, but I am learning to identify the ones that should be acted upon—those which if unfulfilled will vaporize into latent regret. Galápagos was one of those.

Had I not felt so powerfully in the first place, my quest might have faded. But something about my intensity fired my children and kept it going. It might have smoldered for twenty more years had I not expressed it almost flippantly in that auspicious e-mail: "Honey, I need to get away from all

Epilogue: Full Circle

this. I think it's time for the Galápagos." And if that e-mail hadn't arrived at a moment when Heather also felt restless and overworked, I would probably have sat at my computer during those five days in August 2000, tapping out the latest in a series of grant applications—tapping interminably while thousands of miles away on a glassy, sun-drenched sea, Pearl and Henry, Bill, David and the others boarded the *Beluga* without me.

When friends ask if I want to go back, I say, "Of course," and feel perfectly solid in my response. I still yearn to see penguins nesting on Floreana, flightless cormorants holding out vestigial wings to dry as they step from rock to rock on Isabela and Fernandina, and the cartoon-awkward landings of waved albatross skidding onto "runways" in Española. But the other day when a friend asked when I planned to return, I stumbled like a politician unsure of the answer and just wanting the topic to fade.

I remember my daughter's question of long ago, the question that ignited my dream. And I tremble, just as I did before, but for different reasons. I'm no longer afraid of the journey. It's the islands themselves that make me tremble today. Too many people and too many invasive species threaten their very existence. The Galápagos are now on the list of World Heritage Sites in danger, and their preservation has become an Ecuadorian national priority.

The more I think about returning, the more uncertain I become. I know I want others to see what I saw. How else will they understand their beauty and importance? But I favor stricter regulations and higher entrance fees to help safeguard the islands and their dwellers from those who seek to do them harm, or who might haplessly endanger them. Now, with so many problems unresolved, I will stay informed, donate toward their conservation, and spread the word about their fragility.

Since the Galápagos, I have explored Utah's Capitol Reef and Zion national parks with my daughters and traveled with Yolis to Nevada's Red Rock Canyon. Each time I release myself into nature, I feel splashes of wanderlust and a desire to do more and learn more. I may not be a daring adventurer who parachutes from planes, rappels rocky cliffs, or kayaks on the open sea, but I have dreams still unrealized.

At this point I'm unsure where these feelings will lead—perhaps to mastering Spanish and retracing my mother's roots in Chihuahua. Or going to the island of Kythera where my Greek mother-in-law was born. Or finishing, at long last, several books of promising short stories. Or traveling to Darwin's home in Downe to experience the man who inspired my island odyssey.

I think of my sister-in-law Connie and her "might-have-beens," then close my eyes and hear my daughter's words: *"What are you waiting for? Why not now?"*

Chronicle of the Enchanted Isles

Approximately six hundred miles off the Ecuadorian coast, you'll find the Galápagos Archipelago—a renowned chain of thirteen "large" volcanic islands, six smaller ones, and more than a hundred islets and rocks. Across the ancient lava, unique species of giant tortoises for which the islands were named, land and marine iguanas, lava lizards, geckos, and snakes leave their imprint. Mammals like the Galápagos sea lion, fur seal, and rice rat can be found here and nowhere else.

At least forty-one unique species of fish swim the waters of these islands. Seabirds like the Galápagos penguin, petrel, lava gull and flightless cormorants are peerless among the world's creatures. Among the land birds exclusive to these islands are thirteen species of Darwin's finches, four mockingbird species, the Galápagos dove, barn owl, and hawk. One-of-a-kind cotton and tomato specimens, along with endemic plants like the *Scalesia* or giant daisy trees, *Opuntia* cacti, and pink-flowering *Miconia* take their nutrients from the volcanic soil.

In the years since Darwin's visit in 1835, these once pristine islands have become a World Heritage Site in Danger. Introduced species, climate change, human contact, and habitat destruction now place most of the species listed above (and *many* others) in peril, and have resulted in their designation as "critically endangered," "endangered," "vulnerable" or "near threatened."

The following is a quick look at how these islands were formed, populated, and protected as a world treasure:

- Millions of years ago, hot spots between the inner core and outer crust of the earth began to ignite. Columns of hot rock formed beneath the sea where the Nazca, Cocos, and Pacific tectonic plates converge.
- Over hundreds of thousands of years, veins or channels of melted rock called *magma* erupted and forced their way to the surface. The magma created volcanoes.
- About three million years ago, the currently visible volcanic islands began surfacing—first the southeastern most Española-- followed by others until the youngest and western-most islands, Fernandina and Isabella, emerged.
- In the 1400s the Incas discovered the islands. Historians believe they may have been the first humans on the islands.
- In 1535, Fray Tomás de Berlanga, the Bishop of Panama, "officially" discovered the islands which were named *Insulae de los Galápagos* or the "Islands of the Tortoises." His lost ship bound for Peru diverted its voyagers onto the islands in a vain search for water. In the years that followed, pirates, buccaneers, whalers and sealers left indelible human marks on the islands.

Chronicle of the Enchanted Isles

- In 1546, Spanish soldiers with little sailing experience tried to dock on the fog-enveloped Galápagos. They named the mysterious islands *Las Islas Encantadas* or "The Enchanted Islands" because they seemed to float away, making landing impossible.
- Around 1793, British Captain James Colnett explored the whale oil potential of the islands. Thus began the environmentally devastating whaling period and the exploitation of sea lions and tortoises.
- During the War of 1812, U.S. Navy Captain David Porter made detailed observations on the volcanic and natural history of the islands.
- In 1832, Ecuador annexed the islands and began colonization. A penal colony on Floreana Island became the scene of murder and slavery.
- In 1835, Charles Darwin, a young naturalist on the *H.M.S. Beagle*, spent five weeks visiting four islands and collecting specimens that ultimately lead to the popular acceptance of evolution.
- In 1854, *Moby Dick* author Herman Melville visited the islands that he called "five-and-twenty heaps of cinders."
- In 1959, Ecuador declared ninety-seven percent of the land a national park, and the Charles Darwin Foundation for the Galápagos Islands began.
- In 1968, Ecuador employed the first two park rangers.
- By 1970, the islands had become an important tourist attraction.
- In 1978, the Galápagos were among the first three natural World Heritage Sites named by the

United Nations Education, Scientific and Cultural Organization (UNESCO).
- In 2007, approximately thirty thousand people resided on the islands compared to one thousand residents in 1978.
- In 2007, approximately one hundred fifty thousand visited the islands compared to nine thousand tourists in 1978.
- In 2007, UNESCO designated the islands as a World Heritage Site in Danger. During the same year, the president of Ecuador decreed that the conservation of the Galápagos has become a national priority.

Discovering Charles Darwin

When I returned from the Galápagos, I was filled with wonder about the man who plucked these tiny islands from obscurity and propelled them into the headlines. The more I read, the more I wanted to know about Darwin's place in history and his exact contributions.

Born in Shrewsbury, Shropshire, England, on February 12, 1809 (the same day and year as Abraham Lincoln), Charles Darwin lived through dynamic times. When he was six years old, Napoleon I lost his struggle for world power. As Darwin grew into adulthood, the British Empire, under Queen Victoria, became the most powerful and largest political entity in the history of the world—covering roughly one-quarter of the earth's surface. During his lifetime, the United States waged Civil War; John Wilkes Booth assassinated Lincoln; Karl Marx published his Communist Manifesto; serfdom was banned in Russia; and slavery was legally abolished by his own country and by the United States.

Darwin's name is almost synonymous with "evolution," yet he did not originate this concept. Many "evolutionists" came before him—including his own paternal grandfather, Erasmus Darwin (1731–1802), and Jean-Baptiste Lamarck (1744–1829) who, as Charles wrote, "was the first man whose conclusions on the subject excited much attention." In Darwin's time, evolution was not widely accepted because no one had a scientifically defensible explanation for how organisms change over time. Ignited with curiosity, keen observation, and a talent for picking the perfect people to support him, he discovered a key mechanism for how this change takes place—a process he called "natural selection."

In my discoveries about Charles Darwin I was surprised to find that none of his elementary school teachers saw anything exceptional in him. Though he was not an ordinary boy, he never excelled in his early school years—probably because the people and the world outside his intellectually pastel Shrewsbury classrooms were so vivid and stimulating.

The youngest of five, he lost his mother, Susannah Wedgwood, when he was eight. To Robert, his father, the young Darwin was an unmotivated, lazy attention-getter whose greatest potential was to bring shame to the family name. In his autobiography, Charles admitted to being "much given to inventing deliberate falsehoods." He recalled gathering fruit from his father's orchard, hiding it in shrubbery, and then rushing "to spread the news" that he had "discovered a hoard of stolen fruit."

In lieu of schoolwork, Charles spent hours in his father's gardens germinating his naturalist inclinations. His brother, Erasmus, added to his fascination with science by building a small chemistry lab in their backyard shed. Erasmus experimented ardently with Charles—mixing chemicals and producing gasses that sister Caroline feared would someday blow up their house. Working with his older brother, sometimes late into the evening, thirteen-year-old Darwin learned basic methods of scientific experimentation.

The young Darwin read Shakespeare, Byron, and Scott extensively, and re-read a book entitled *Wonders of the World* many times. He collected rocks, insects, coins and stamps;

and spent summers hiking the backwoods of northern Wales with his family.

Charles came from a wealthy family in which male family members excelled. His father was a physician; his father's father a physician, naturalist, and poet; and his maternal grandfather, Josiah Wedgwood, a renowned potter who patented Wedgwood pottery and china. Robert Darwin was intent on securing his son a successful future, and so he encouraged, cajoled, and pushed him to excel. Finding Charles unmotivated and mediocre in his classical studies, he withdrew him from Reverend Samuel Butler's boys' boarding school in Shrewsbury and enrolled him, at age sixteen, in the University of Edinburgh—famous for its medical school. Though bored in his classes, and showing no interest in a medical career (to the dismay of his father), Darwin began a pattern of choosing mentors who would educate and support him throughout his life.

At age seventeen, Charles sought out his Edinburgh neighbor John Edmonstone, a freed black slave from Guyanna, South America. He paid Edmonstone, who taught taxidermy to medical students, for private lessons on stuffing animals. His friendship with this mentor brought him an invaluable skill. It also fired him with a disdain for slavery and a strong desire to see the tropical rain forests of South America. At this time, Darwin also aligned himself with Robert Grant, a zoology professor and radical promoter of evolution; and with William Macgillivray, the curator of the Edinburgh's Natural History Museum. Macgillivray taught him animal anatomy and botany and convinced him to begin making scientific notes on his observations. Growing impatient with his son's lack of focus on a medical career, his father withdrew him from the University of Edinburgh and enrolled him at Cambridge to prepare for a career in the ministry. But he was too late. Darwin was well on his way to becoming a naturalist.

At Cambridge University, instead of focusing on the ministry, another career that didn't interest him, twenty-year old Darwin partied a lot and surrounded himself with science advocates and experts. A cousin interested him in collecting beetles, an avocation he so zealously adopted that he spent more time with his beetles than with his studies. Even beautiful women couldn't compete with Darwin's bugs. "I will give a proof of my zeal," he wrote.

> One day on tearing off some old bark, I saw two rare beetles and seized one in each hand; then I saw a third and new kind, which I could not bear to lose, so that I popped the one which I held in my right hand into my mouth. Alas it ejected some intensely acrid fluid, which burnt my tongue so that I was forced to spit the beetle out, which was lost, as well as the third one.

Cambridge naturalist instructor Reverend John Henslow became one of Charles' greatest mentors and friends—expanding his young protégé's knowledge of botany, chemistry, geology, entomology, mathematics, and mineralogy—and ultimately getting him a position as naturalist aboard the *H.M.S. Beagle*. To further his studies, Darwin read books on scientific investigation and on South America. In 1831, at age twenty-two, he graduated from Cambridge with a Bachelor of Arts degree—his science education having been learned "informally."

Fascinated with geology, he returned to Cambridge and audited lectures given by the renowned professor Adam Sedgwick. Sedgwick respected the young man and invited him to be his assistant on a geology field tour of northern Wales. Armed with new skills and knowledge, Darwin was ready for a great adventure that would test all he'd ever learned.

The white-bearded Englishman most of us picture as Charles Darwin was only twenty-two when he set sail as a clean-shaven naturalist on the *H.M.S. Beagle*. The primary purpose for the voyage was not to make scientific history, but to produce accurate charts and maps of South America and further England's trade interests with this resource-rich land. In addition to charting, the *Beagle's* captain, Robert Fitzroy, had an avid interest in natural science.

In the summer of 1831, Darwin's friend and botany mentor Reverend John Henslow heard that Captain Fitzroy needed a suitable person to collect specimens throughout the journey. Henslow wrote Darwin about this exciting opportunity, adding that the captain wanted much more than a gifted naturalist. He wanted a companion who was a true gentleman. Henslow recommended Darwin for the post. Fitzroy agreed, and Charles, who had long dreamed of going to South America, enthusiastically accepted the invitation.

When his father and sisters learned of his impending voyage, they convinced him that this adventure was just another scheme to distract him from his focus on the ministry. Loyal to his family, Charles declined the invitation. Seeing his son's disappointment, Robert Darwin opened what he probably thought was an unlikely window. "If you can find any man of common-sense who advises you to go," he told his son, "I will give my consent." That man turned out to be Darwin's Uncle Josiah Wedgwood, who convinced Robert that this journey would become a milestone in the young man's career.

With his father's approval, Darwin went to see Fitzroy in London where he learned that Fitzroy had offered the position to another man who had declined it. Believing that the shape of Darwin's nose indicated poor stamina, the captain almost disqualified him, then decided to take a chance on the enthusiastic applicant. After months of delay, the *Beagle* set sail out of Plymouth Harbor on December 27, 1831.

"The voyage of the Beagle," Darwin wrote years later, "has been by far the most important event in my life, and has determined my whole career." Seasick throughout the almost-five-year journey, he maintained his stamina. Peace loving and unaccustomed to life at sea, he endured ocean storms, earthquakes, native warfare, volcanic eruptions, and disturbing arguments with the captain over slavery and Darwin's "infamous line of thinking." They sailed and docked along both coasts of South America and ventured into Tierra del Fuego, the Falkland Islands, Australia, New Zealand, and the Galápagos Archipelagos.

On land, free from the rolling motion of the sea, Darwin concentrated on observing, collecting specimens, and questioning everything he saw. Along the way he used his taxidermy skills and stuffed animals, then mailed boxes of fish, reptiles, rocks, birds, insects, and fossils to Reverend Henslow, who distributed the precious cargo among the best natural scientists in England. In October 1833 while on shore collecting, Darwin (who was called "Philos" by Fitzroy) received the following message from the captain:

> My dear Darwin [:] Two hours since, I received your epistle, dated 26th and most punctually and immediately am about to answer your queries. (*mirabile!!*) But firstly of the first—my good Philos why have you told me nothing of your hairbreadth escapes & moving accidents[.] How many times did you flee from the Indians?

How many precipices did you fall over? How many bogs did you fall into? –How often were you carried away by the floods? and how many times were you kilt?—that you were not kilt dead I have visible evidence in your handwriting....

Throughout South America, Darwin collected so many fossils the captain objected to his bringing too much junk aboard. The collection job became so overwhelming that Darwin's father, who had come to respect his son's achievements, paid the salary for crewman Syms Covington to be Charles' servant. Syms quickly learned taxidermy and became a valuable asset to the ardent explorer.

Wherever they docked, Charles scrutinized what others before him had seen and passed over. Why, he puzzled, were there no animals in South America that resembled the huge fossils he collected? Had the environment changed dramatically, robbing them of their sustenance, or was something else responsible? Why did the walls of the Rio Santa Cruz River Valley and the Andes Mountains themselves contain layers of seashells? Had the land thrust upwards from below the sea? What caused the change of species in plants and animals from one geographical location to another?

When the twenty-six-year-old Darwin reached the Galápagos in 1835 and collected his now-famous finches, his developing belief in natural selection had begun to solidify. The evidence supporting an ever-changing planet was mounting—contrary to the powerful religious belief that God's creations were perfect from the beginning and needed no alterations. In his autobiography years later, Darwin wrote that the "supposition that species gradually became modified...haunted me."

By the time he was twenty-seven, about a year after his return from the *Beagle* expedition, Darwin believed firmly in "transmutation"—the theory that animals adapted to new circumstances, evolving over much time into new species. He had read the French scientist Lamarck's and his grandfather Erasmus' works on the theory. Transmutation, he concluded, followed strict natural laws and invalidated the popular belief held by many of his peers—that God alone, acting without the mechanism of evolution, directly created each new species. Because he was not ready to confront the traditional establishment, he confided in no one but his brother Erasmus.

As his theories fermented, he continued to pick his mentors and colleagues well.

London Zoo Museum ornithologist John Gould gave Darwin an exciting new perspective about the bird specimens he collected in the Galápagos. They were not, as Darwin supposed, finches, gross beaks, wrens, and blackbirds. All were finches that varied mainly in the shapes of their beaks. Darwin's job was to figure out how the variations of the finch from island to island could have occurred.

His poor job of labeling the specimens turned out to be a stumbling block. Because he had not identified the birds by island, he consulted with fellow *Beagle* crewmen who had collected similar specimens and labeled their locations. He then proceeded to re-examine the finches, this time with fresh insight. Finches with large nuts as their primary food source had powerful beaks to break them. Those that lived where small nuts and seeds were plentiful had beaks suitable for cracking. Those surrounded by fruit-bearing plants had parrot-shaped beaks, while those dependent on insects had slender bills. Each species, he concluded, must perform a unique function on its own island.

At one point, concerned about the possible effects of going public with his theories, Darwin dreamt he was hanged for his heretical beliefs. At age twenty-eight, he began to endure mysterious health disorders that persisted for the remaining forty-five years of his life. He continued his meticulous work in spite of debilitating stomach pains, nausea, vertigo, vomiting, migraines, and heart problems. The cause of his illness is still not known for certain. Darwin may have attributed his health problems to stress, a diagnosis that many experts today who specialize in psychosomatic disorders support. Other experts believe he may have suffered from one or a combination of problems, including anxiety, lactose intolerance, arsenic poisoning, multiple allergies, or Chagas' Disease caused by the bite of a *Triatima infestans* or the "assassin bug" while he was at the foot of the Argentine Andes.

In his *Beagle* journal, Darwin described the bite:

> At night I experienced an attack (for it deserves no less a name) of the Benchua, a species of *Reduvius*, the great black bug of the Pampas. It is most disgusting to feel soft wingless insects, about an inch long, crawling over one's body. Before sucking they are quite thin but afterwards they become round and bloated with blood.

The year 1838 turned out to be a watershed year for Darwin, both professionally and personally. In his autobiography written years later, he recalled reading "for amusement" a paper by Thomas Robert Malthus, a political economist. In the "Essay on the Principle of Population" published in 1798,

Malthus observed that when plants and animals produce more young than can survive, they compete for survival. Darwin wrote of the epiphany he experienced when reading Malthus:

> ...it at once struck me that under these circumstances favourable variations would tend to be preserved, and unfavourable ones to be destroyed. The result of this would be the formation of new species. Here, then, I had at last got a theory by which to work.

In 1838, Charles also began courting Emma Wedgwood, his first cousin, who was the daughter of his staunch advocate Uncle Josiah. His father warned him not to mention his revolutionary ideas to the somewhat traditional Emma. But Darwin knew he could not build a lasting relationship on half-truths. Emma must understand the potential controversies that lay ahead for them if she consented to be his wife, and so he confided in her. Nine days before their wedding, he wrote her: "I think you will humanize me, and soon teach me there is greater happiness than building theories and accumulating facts in silence and solitude." On January 29, 1839, close to his thirtieth birthday, they exchanged vows at St. Peter's Church at Maer. Throughout their thirty-four year marriage, he continued with his honesty, and she with unflinching loyalty.

Charles and Emma had ten children—seven who lived into adulthood. Because they were wealthy, Charles didn't worry about supporting his family. He spent his time researching, writing, reading, raising orchids and pigeons, and being with his wife and children whom he adored. Weakened by chronic health problems, Darwin rarely traveled. However, "Down House," his woodsy country estate in Kent, became a meeting place and research laboratory for eminent scientists of the day.

In 1842, a month after they moved from London to Down House, Charles and Emma lost their infant daughter, Mary Eleanor. Grieving, he sought escape in writing about the geology of the Galápagos. In 1844, eight years after his *Beagle* voyage, Darwin risked discussing his writings on *transmutation*, or how species change into other species, with eminent botanist Joseph Hooker. Though skeptical, Hooker, a trusted friend, became his confidant and collected related readings for Darwin to study. About this time, a controversial book on the natural history of creation introduced the concept of transmutation to the general public. Darwin could tell by the furor the book raised among his peers that his own writings would face harsh scrutiny. For his work to be taken seriously, he would have to establish himself as an expert on changing species, and so, for the next eight years, he devoted himself to much-needed research on barnacle specimens collected during the *Beagle* expedition.

By the time Darwin was thirty-eight, he suffered continuously from stomach convulsions, twitching, and dizziness. His writing and research on barnacles became increasingly difficult, and he was sure he was going to die. At age forty—ready to try anything, he began a ritual of ice-cold-water treatments that brought him intermittent relief over the years.

In 1851, when he was forty-two, Darwin's favorite child, his ten-year-old daughter, Annie, died of what was probably tuberculosis. Annie was an affectionate little girl with a special love for her father. Her death so stunned him that his fears about offending the scientific and church establishments paled. "We have lost the joy of the household," he wrote, "and the solace of our old age: she must have known how much we loved her... and shall ever love her dear joyous face. Blessings on her."

After Annie's death, Darwin threw himself all the more fervently into his barnacle research. Two years later, he received the highest scientific honor the Royal Society can bestow—for his barnacle research, still in progress, and for his *Beagle* geology discoveries. A year later, with his barnacle research published, Darwin returned to his writings on transmutation and began raising pigeons and studying how nature transports seeds. Adding to his already-vast knowledge, he discovered that species can move from one environment to another by floating or being blown by the wind.

In 1856, at age forty-seven, he began writing his theory of natural selection to explain *how* animals and plants split off into separate species. He concluded that within species, individuals vary. Some are larger, stronger, have different markings, longer legs, thicker fur, or a unique body shape. When natural differences increase their odds for survival, their offspring survive in greater numbers and pass on their advantages to succeeding generations.

In general, he said: "It's not the strongest of the species that survive, nor the most intelligent, but the one most receptive to change." Thus, nature selects how new species are to evolve. Ironically, as Darwin drafted his ideas in England, a colleague mailed him a paper on species variations written the previous year by thirty-two-year-old naturalist Alfred Russel Wallace who was researching the Malay Peninsula. Darwin wrote to congratulate Wallace, and the two continued to correspond concerning their mutual interests.

In 1857, Darwin sent a letter to Harvard natural history professor Asa Gray, including an abstract of his upcoming book on natural selection. In 1858, Wallace sent Darwin his completed, unpublished paper on natural selection—a paper containing conclusions dramatically similar to the ones Darwin had been working on for more than twenty years. The paper couldn't have come at a worse time. Distraught over his critically ill baby son, Darwin now had to face the possibility that all he had worked toward was unattainable. He confided his dilemma to his geologist-friend Charles Lyell. Lyell conferred with Hooker who agreed that both Darwin's and Wallace's writings should be read at the upcoming meeting of the Linnean Society—a distinguished life science organization.

At first, Darwin refused to have his own writings presented along with Wallace's. Not knowing the benevolent disposition of his fellow naturalist, Darwin thought that Wallace would consider this "unjustifiable." Later he discovered how "generous and noble" Wallace was. Darwin was also dissatisfied with the quality of his own writing. "The extract from my MS. and the letter to Asa Gray," he wrote, "had neither been intended for publication, and were badly written. Mr. Wallace's essay, on the other hand, was admirably expressed and quite clear."

On June 28, 1858, Darwin's eighteen-month-old baby, Charles, died of scarlet fever. On July 1, Darwin's friends Joseph Hooker and Charles Lyle presented both his and Wallace's papers at the end of a tediously long Linnean Society agenda. Neither author was physically present in the hall that day (Darwin at home grieving and Wallace in Malaysia), and the tired membership, anxious for the meeting to end, gave little notice to what should have been a dramatic moment in history. Leaving the hall, Linnean members went about their normal lives, not suspecting that what they had just heard would ultimately become the talk of the scientific community. Natural selection would no longer be the private theory of two independent thinkers. It would become a public matter.

Much has been written about why Darwin and not Wallace became synonymous with evolution. Wallace appears to have allowed Darwin the spotlight—perhaps because he respected the many years Darwin had devoted to the topic. Wallace himself went on to become a notable naturalist and expert on the distribution of animal species throughout the world. The two men continued to communicate until Darwin's death in 1882.

Eighteen months after the public reading, Darwin published his masterpiece—in great part inspired by his five weeks on the Galápagos Islands. *The Origin of Species by Means of Natural Selection: Or the Preservation of Favored Races in the Struggle for Life* rapidly received praise and damnation from many quarters. The waves of controversy that surrounded it caused Darwin to lose the opportunity for knighthood. Although he never publicly defended his book nor his ideas, his many loyal friends took on his enemies.

Darwin himself, though weakened by continuous bouts of illness, continued his research and writing. His curiosity was so intense and his powers of observation so keen that he wasn't content to putter in his garden or raise animals for pure amusement. Wherever his interests led him, he meticulously studied, measured, analyzed and was driven to find answers to questions most humans never ponder: Questions like: How do the many different species of orchids attract bees and moths that are perfectly equipped to pollinate their unique petal formations? How similar

Charles Darwin

are humans and animals in their expressions of joy, grief, fear, anger and other emotions? What are the differences in plants that pollinate themselves and those that are pollinated by insects or birds?

Working at home, he was free to take daily walks through the woods and was never without the support of Emma and their seven children. A patient father and loving husband, he enjoyed the antics of his children and spent hours playing backgammon and listening to the novels Emma read to him daily.

On April 15, 1882, at age 73, Charles Darwin died at Down House with his family at his bedside. He was buried in Westminster Abbey close to Sir Isaac Newton. In addition to his works on the *Beagle* exploration and natural selection, he had written or edited more than twenty books and one hundred fifty articles on topics such as climbing and insectivorous plants, orchids, barnacles, worms, breeding, coral reefs, pigeons, and on his grandfather's and his own life stories.

Discovering Charles Darwin has been a *Galápagoan* experience for me. As with my own voyage to *Las Islas Encantadas*, I started with great expectations and little knowledge. I ended with a collection of fermenting images and a burning desire for more. The Darwin I have come to love was generally unimpressed with himself. "My power to follow a long and purely abstract train of thought is very limited," he wrote. He seemed in awe of what a person with "moderate abilities" could accomplish. "It is truly surprising," he said, "that thus I should have influenced to a considerable extent the beliefs of scientific men on some important points."

Darwin took credit for his "methodical" habits but doubted that he could have succeeded in his quest without the leisure time his wealthy lifestyle provided and his ill health demanded. His greatest qualities, he believed, were his "love of science," "unbounded patience in long reflecting over any subject," and his "industry in observing and collecting facts." He also admitted to "a fair share of invention."

Discovering Charles Darwin

Ordinary people observe their surroundings and move on, drawing few conclusions, and ultimately forgetting what they've seen. Charles Darwin, in spite of his own protestations to the contrary, was not ordinary. He was a man with incredible tenacity and integrity—a man who persisted with his revolutionary thinking and painstaking documentation in spite of constant debilitating illness, the devastating loss of three children, and pressure to keep his controversial ideas to himself. When his father warned him not to tell Emma about his controversial beliefs, he risked her love by telling her the truth. At a time when both England and the U.S were profiting from human abuse, he spoke out against slavery. "It makes one's blood boil, yet heart tremble," he wrote, "to think that we Englishmen and our American descendents, with their boastful cry of liberty, have been and are so guilty...." Upon receiving Alfred Russel Wallace's paper, he could have hurriedly issued his own paper and taken the full credit for their mutual discovery. His ethics, however, made such an act unthinkable.

When I think of Darwin today, I look far beyond the picture of a stately, white-bearded old man. Instead, I see a distracted young man daydreaming in his university classrooms, a naturalist battling seasickness while passionately collecting specimens that will ultimately document the ever-evolving face of our planet. I see a grieving father, a husband listening to his wife read the classics, a gardener touching the silky petals of orchids in his green house. I picture a man tormented by a mysterious illness writing at his desk with children romping about, a scientist studying his beloved pigeons and probing unparalleled theories with colleagues in his home.

I think also of what renowned *New York Times* science journalist Walter Sullivan wrote in his 1972 introduction to *Charles Darwin: The Voyage of the Beagle*. It was Darwin, he said, "who read nature's message: that we as men are not isolated from nature; that we are, indeed, a part of it." Throughout his adult life, Darwin maintained a profound respect for nature's power to create and destroy and for humanity's role in both the creation and the destruction. "His lesson should not be lost on us today," pleaded Sullivan. "For our works have now begun so to overwhelm the environment that we can only survive if we learn not only to dominate, but to rule wisely."

Bibliography

Galápagos and Quito

Boyce, Barry. *A Traveler's Guide to the Galápagos Islands.* 3rd ed. Aptos, CA: Galapagos Travel, 1998.

Charles Darwin Foundation and the Charles Darwin Research Station. 2006. 28 November 2008 <http://www.darwinfoundation.org/>.

Diamond, Antony W. and Elizabeth A. Schreiber. "Magnificent Frigatebird." *The Birds of North America Online.* <http://bna.birds.cornell.edu/bna/species/601/articles/introduction>.

Droste, Bernd von. "Global Galapagos." *Galapagos News.* Galapagos Conservancy. Spring/Summer 2008 <http://www.galapagos.org/2008/index.php?id+88/>.

Galapagos Conservancy. 28 Nov. 2008 <http://www.galapagos.org/>.

Galapagos Conservation Trust. 28 Nov. 2008 <http://www.gct.org/>.

Galapagos Geology on the Web. Ed. William M. White. 2001. Cornell University Department of Geological Sciences. 29 Sept. 2008 <http//www.geo.cornell.edu/geology/Galapagos.html>.

Global Volcanism Program. Smithsonian National Museum of Natural History. 25 June 2003 <http://www.volcano.si.edu/>.

Gorry, Conner. *Read This First: Central & South America.* Victoria: Lonely Planet, 2000.

Gray, Louise. "Lonesome George's First Sex in Decades Ends in Disappointment." Telegraph.co.uk. December 6, 2008. http://www.telegraph.co.uk/earth/wildlife/3566021/Lonesome-Georges-first-sex-in-decades-ends-in-disappointment.html>.

Greenwood, William and Norton, Robert L. "Novel Feeding Technique of the Woodpecker Finch." *Journal of Field Ornithology* 70 (1999): 104-106.

Hofkin, Bruce et al. "Ancient DNA Gives Green Light to Galápagos Land Iguana." *Conservation Genetics* 4 (2003): 105–108.

Horwell, David and Pete Oxford. *Galápagos Wildlife: Visitor's Guide.* Chalfont St. Peter, Bucks: Bradt Publications; Old Saybrook: The Globe Pequot Press Inc., 1999.

Kosseff, Lauren. "Woodpecker Finches Use Improvised Beaks." *Animal Cognition Web Site.* Tufts University, 2005 <http://www.pigeon.psy.tufts.edu/psych26/birds.htm#finch>.

The Marine Mammal Center. "The Pinnipeds: Seals, Sea Lions, and Walruses." <http://www.marinemammalcenter.org/learning/education/pinnipeds/pinnipeds.asp>.

National Science Teachers Association. "Letters from the Galapagos." <http://www.nsta.org/publications/interactive/galapagos/activities/pdf/letters.pdf>.

Proaño, María and Bruce Epler. 2007. "El turismo en Galápagos: una tendencia al crecimiento." *En: Informe Galápagos* 2006-2007. Parque Nacional Galápagos, Fundación Charles Darwin, Instituto Nacional Galápagos, Puerto Ayora, Galápagos, Ecuador.

Pululahua Geobotanical Reserve. 2006-2008. Pululahua Hostel. 10 Jan. 2009. <http://www.pululahuahostal.com/html/pululahua_geobotanical_reserve.html>.

Rothman, Robert. Galapagos. 23 June 2003 <http://people.rit.edu/rhrsbi/GalapagosPages/NewGalapagos.html>.

Roy, Tui De. *Spectacular Galapagos: Exploring an Extraordinary World*. Hong Kong: Hugh Lauter Levin Associates, Inc., 1999.

Snell, Howard. Interview. Ask the Scientists. Public Broadcasting System. 1997. <http//www.pbs.org/safarchive/3_ask/archive/bio/101_snell_bio.html>.

Tebbich, Sabine et al. "Cognitive Abilities Related to Tool Use in the Woodpecker Finch." *Animal Behaviour* 67 (2004): 689-697.

Volcano Discovery. "Latest Volcanic News from the Galapagos Islands (Ecuador) May 15, 2005-June, 07, 2008."<http:www.volcanodiscovery/volcano-tours/volcanoes/pacific/galapagos.html>.

Volcano World. Ed. Oregon State University Department of Geosciences. 1 October 2003 <http://volcano.oregonstate.edu/>.

Watkins, Graham and Felipe Cruz. (2007)."Galapagos at Risk: A Socioeconomic Analysis of the Situation in the Archipelago." Puerto Ayora, Province of Galapagos, Ecuador, Charles Darwin Foundation.

Weiner, Jonathan. *The Beak of the Finch*. New York: Alfred Knopf, 1994.

Woram, John. "The First Iguana Transfer." *Noticias de Galápagos* 51 (1992): 22.

Woram, John. "Who Killed the Iguanas?" *Noticias de Galápagos* 50 (1991): 12-17.

Darwin Bibliography

Archibald, J. D. 2008. "Edward Hitchcock's pre-Darwin (1840) 'Tree of life.'" *Journal of the History of Biology.* Online First. doi:10.1007/s10739-008-9163-y. <http://www.springerlink.com/content/a647j5583641537v/>.

Burkhardt et al. eds. *The Correspondence of Charles Darwin Volume I 1821-1836.* Cambridge University Press, 1985.

The Complete Works of Charles Darwin Online. John van Wyhe. 22 Jan. 2003. <http://darwin-online.org.uk/majorworks.html>.

Charles Darwin and His Writings. *The C. Warren Irvin, Jr. Collection of Charles Darwin and Darwinia.* Department of Rare Books & Special Collections. University of South Carolina. 7 July 2003 <http://www.sc.edu/library/spcoll/nathist/darwin/darwin.html>.

Darwin, Charles. *The Autobiography of Charles Darwin, 1809-1882.* Ed. Nora Barlow. New York: W.W. Norton & Co., 1958.

Darwin, Charles. *The Voyage of the Beagle.* New York: Mentor-Penguin Group, 1988.

Darwin Correspondence Project. Cambridge University Library. <http://www.darwinproject.ac.uk/>.

Evolution Library. 2008. Public Broadcasting System. 25 Sept. 2008. <http://www.pbs.org/wgbh/evolution/library/faq/>.

Keynes, Randal. *Darwin, His Daughter and Human Evolution.* New York. Riverhead, 2002.

Leff, David. *AboutDarwin.Com.* 10 Feb. 2008 <http://www.aboutdarwin.com>.

Bibliography

Nichols, Peter. *Evolution's Captain: The Story of the Kidnapping that Led to Charles Darwin's Voyage Aboard the Beagle.* New York. Harper Perennial, 2004.

Pasnau, Robert O. "Darwin's Illness: A Biopsychosocial Perspective." *Psychosomatics* 31 (1990): 121-128.

Reveal, James L., Paul J. Bottino, and Charles F. Delwiche. *The Darwin-Wallace 1858 Evolution Paper.* University of Maryland. 6 Jan. 2003 <http://www.plantsystematics.org/reveal/PBIO/darwin/darwindex.html>.

Sato, Akie, et al. "Phylogeny of Darwin's Finches As Revealed by mtDNA Sequences." *Proceedings of the National Academy of Sciences of the United States of America*, v. 96 (9), 1999. <http://www.pnas.org/content/96/9/5101.full?ck=nck>.

Smith, Charles H. *The Alfred Russel Wallace Page.* 6 Jan. 2003 <http://www.wku.edu/~smithch/index1.htm>.

Sullivan, Walter. Introduction. *The Voyage of the Beagle.* By Charles Darwin. New York: Mentor, 1988. vii-xviii.

The TalkOrigins Archive: Exploring the Creation/Evolution Controversy. 29 Sept. 2008. <http://www.talkorigins.org>.

University of California Museum of Paleontology. Exhibit Halls: "Welcome to the Evolution Wing." University of California at Berkeley. Sept. 2008. <http://www.ucmp.berkeley.edu/history/evolution.html>.

Urbanowicz, Charles F. "Teaching as Theatre: Some Classroom Ideas, Specifically Those Concerning Charles R. Darwin (1809-1882)." 23 Jan. 2003 <http://www.csuchico.edu/~curbanowicz/Darwin2000.html>.

Urbanowicz, Charles F. "Urbanowicz on Darwin." 29 Sept. 2003 <http://www.csuchico.edu/~curban/Darwin/DarwinSem-S95.html>.